Sustainable Tourism

Thomas Walker • Ender Demir
Gabrielle Machnik-Kekesi • Victoria Kelly
Editors

Sustainable Tourism

Frameworks, Practices, and Innovative Solutions

palgrave
macmillan

Editors
Thomas Walker
Concordia University
Montreal, QC, Canada

Ender Demir
Reykjavík University
Reykjavik, Iceland

Gabrielle Machnik-Kekesi
Concordia University
Montreal, QC, Canada

Victoria Kelly
Concordia University
Montreal, QC, Canada

ISBN 978-3-031-43527-0 ISBN 978-3-031-43528-7 (eBook)
https://doi.org/10.1007/978-3-031-43528-7

This Palgrave Macmillan imprint is published by the registered company Springer Nature Switzerland AG.
The registered company address is: Gewerbestrasse 11, 6330 Cham, Switzerland

Paper in this product is recyclable.

Acknowledgements

We acknowledge the financial support provided through the Jacques Ménard—BMO Centre for Capital Markets. We appreciate the excellent copy-editing and editorial assistance we received from Charlotte Frank, Meaghan Landrigan-Buttle, Miles Murphy, and Eimear Rosato.

Contents

Notes on Contributors

Katariina Ala-Rämi is a senior researcher at the University of Oulu, Kerttu Saalasti Institute, MicroENTRE. Her research focuses on the competitiveness of micro-enterprises, networks and environments, regional resilience, and regional development. She is particularly interested in the activities of small enterprises in sparsely populated areas and their impact on the business environment. Since completing her PhD, Ala-Rämi has worked on various research and development projects, with funding from the Academy STN, TEKES, InterregEUROPE, InterregNPA, KareliaCBC, and ERDF/ESF.

Manuel Au-Yong completed his post-doctoral studies at the University of Aveiro and has a PhD in Industrial Engineering and Management from the Faculty of Engineering at the University of Porto, 2012, as well as an MBA degree from Cardiff Business School. Manuel is an assistant professor at the University of Aveiro. He is affiliated with the Research Unit on Governance, Competitiveness and Public Policies (GOVCOPP, full researcher) and the Institute for Systems and Computer Engineering, Technology and Science (INESC TEC, collaborator). He serves as the director of the Master's Degree in Management at the University of Aveiro, serves as a member of the Executive Committee, and coordinates the Committee for Ethics and Social Responsibility of his department—DEGEIT—at the University of Aveiro.

Monika Bachinger received her PhD from the Faculty of Economics at the Catholic University Eichstaett-Ingolstadt, Germany. She is Professor of Tourism at the University of Applied Forest Sciences in Rottenburg, Germany. Her main fields of research are experience design in nature tourism, cultural ecosystem services and tourism product development, and the sustainable management of destinations. She has contributed to approximately 40 publications in the field of entrepreneurship and innovation in tourism, transdisciplinary research, sustainability transformation in tourism, as well as destination management. She is a member of the German Association of Tourism Research (DGT), the Association of Entrepreneurship Research (FGF), and the Working Group on Tourism Research (AKTF) of the German Association of Geographers (DGfG).

Miraj Ahmed Bhuiyan completed his PhD in Industrial Economics. He is working at the School of Economics at Guangdong University of Finance and Economics. His research focuses on the green economy and growth, environmental economics, and sustainable tourism. He teaches international economics, international marketing, cross-cultural business communication, and Chinese business culture courses.

Fatima L. Carvalho holds a PhD from Essex University and is a researcher for CinTurs (the university's Research Centre for Tourism, Sustainability, and Well-being). She lectures on sustainability and governance and is actively engaged in various research projects. She has published her work in books and journals and regularly presents at national/international conferences on themes such as tourism sustainability, governance, citizenship, and educational policy.

Carlos Costa is a full professor at the School of Economics, Management, Industrial Engineering, and Tourism of the University of Aveiro and serves as a tourism expert at the European Commission. He is the editor-in-chief of the *Journal of Tourism & Development* (Revista de Turismo e Desenvolvimento) and associate editor of the *Tourism Review*, the *Journal of Destination Marketing Management*, and the *Journal of Hospitality and Tourism Management*. He holds a PhD and MSc in tourism management (University of Surrey, UK). Carlos is the leader of the PhD Tourism Program and head of the Tourism Research Centre of the University of

Aveiro. He is also the scientific coordinator of the "idtour-unique solutions" tourism start-up company based at the university. He works as a consultant for Portugal's National Agency of Accreditation (A3Es), the National Foundation for Science and Technology (FCT), and for the Cyprus Agency of Quality Assurance and Accreditation in Higher Education (CYQAA).

Tiziana Crovella holds a PhD in Technology and Management (specifically Commodity Science) and is a subject expert and research fellow at the Department of Economics, Management, and Business Law at the University of Bari Aldo Moro, Italy. She holds a Master's Degree in the Economics and Management of Tourism. Her research focuses on the maritime sector and tourism, material flow analysis (MFA), the circular economy, the evolution of packaging, sustainable agriculture, sustainability indicators, and capacity building.

Ender Demir is Associate Professor of Finance at the Department of Business Management, Reykjavík University, Iceland, and acts as the Director of the Corporate Finance Master's Program. Dr. Demir is the conference coordinator of the Eurasia Business and Economic Society (EBES). He is part of the editorial review board for the *Annals of Tourism Research*, serves as Associate Editor for the *Eurasian Business Review*, Subject Editor for the *Journal of Multinational Financial Management*, and as Associate Editor for *Heliyon Business and Economics*. His research has been published in a variety of peer-reviewed international journals, including *Tourism Management*, the *Emerging Markets Review*, *Finance Research Letters*, the *Journal of Financial Stability*, the *Annals of Tourism Research*, and the *Journal of International Financial Markets, Institutions & Money*.

José Luis Durán-Román has a degree in Business Administration and Management and a Master's degree in Economics Research. José Luis is an International Doctor of Economics and Business Sciences and works as an assistant professor in the Department of Business Organization, Marketing, and Sociology at the University of Jaén, Spain. In addition, he is a member of the Laboratory of Analysis and Innovation in Tourism (LAInnTUR) at the University of Jaén. His main research interests focus on tourism economics, the sustainability of tourism, competitiveness, and tourist taxes and he has written extensively in these areas.

Silvia Fernandes is an assistant professor at the Faculty of Economics, University of Algarve. She holds a PhD from the same university and is a member of the Research Centre for Tourism, Sustainability and Well-being (CinTurs). She lectures on Information Technology/Systems and Innovation Management at the undergraduate and master's levels. She has several publications in books and journals and has presented her work at national/international conferences on topics such as enterprise information systems, smart tourism, innovation, and technological diffusion.

Alba García-Milon is a PhD candidate and a researcher at the Faculty of Business Administration in the University of La Rioja (Spain). Her research interests include tourist shopping, consumer behavior, and new technologies adoption. She has presented her work at numerous national and international conferences. She has recently published papers in international journals such as *Telematics and Informatics* and *Tourism Management*.

Yoshihiro Hamaguchi is an assistant professor in the Department of Management Information at the Kyoto College of Economics. His research interests include macroeconomics, environmental economics, the political economy, international economics, and tourism economics. Specifically, his research focuses on the mechanisms and policy instruments of sustainable development based on the theory of endogenous economic growth. His work has been published in refereed international journals such as *Transport Policy*, *Tourism Economics*, and the *Journal of Macroeconomics*. Additionally, he has received awards for the number of citations he has received.

Emma Juaneda-Ayensa holds a PhD in Economics and Management with European Mention and is Professor of Business Organization at the University of La Rioja (Spain). Her main research lines are quality management systems and customer satisfaction, the social economy and sustainable development, and technological factors conditioning user behavior. As a specialist in quality management, from an internal and external perspective, she is interested in topics such as human resources development and consumer behavior. She has participated in numerous conferences nationally and internationally. Her research papers have been published in journals such as the *International Journal of Internet Marketing and Advertising* and *Tourism Management*.

Victoria Kelly holds a BSc in Biology with an additional major in Irish Studies. She has worked with the Emerging Risks Information Center at Concordia University since 2020, where she has been involved in numerous book projects and research papers in the area of sustainability, fintech, and climate change and has assisted with the center's administration. Victoria is pursuing a Master's degree in History at Concordia University, examining the intersection of maternal feminism and eugenics in Canada.

Giovanni Lagioia holds a PhD in Commodity Science and serves as a full professor in the Department of Economics, Management, and Business Law at the University of Bari Aldo Moro, Italy. He is the Dean of the Department of Economics, Management, and Business Law. He teaches courses on commodity science, the commodity science of natural resources, and the management of technological processes and innovation. His scientific research deals with many topics: bioenergy, end-of-life management, material flow analysis (MFA), input-output analysis, and the circular economy.

Yaiza López-Sánchez holds a PhD from the University of Jaén in Social and Legal Sciences, with International and Cum Laude mentions, and obtained an Extraordinary Doctorate Award in the Social Sciences from the University of Jaén. She has completed four postgraduate degrees including a Master's degree in Economics and Territorial Development, a Master's degree in Tourism, Archaeology and Nature, a University Expert degree in the Sustainable Strategic Management of Tourist Destinations, and a University Expert degree in Advanced Methods of Applied Statistics. She is an associate professor in the Department of Economics at the University of Jaén and is a member of both the research team Laboratory of Analysis and Innovation in Tourism (LAInnTUR) and the Chair of Inland Tourism at the University of Jaén. Her research on tourism economics focuses on the sustainability of tourism, the analysis of tourism markets, tourism taxation, tourism competitiveness, and the planning and management of tourism destinations.

Gabrielle Machnik-Kekesi is a research associate at the Emerging Risk Information Center (ERIC) at the John Molson School of Business, Concordia University, Montreal, and a Hardiman Research Scholar (2021–2025) at the University of Galway, Ireland, where she is pursuing her PhD. She holds an Individualized Program Master's degree from Concordia University, Montreal, and a Master's in Information Studies from McGill University, Montreal.

Gianna Moscardo is Professor of Tourism in the College of Business, Law, and Governance at James Cook University (JCU), Australia. Prior to joining JCU, she was the tourism research project leader for the CRC Reef Research. Her qualifications in applied psychology and sociology support her research interests in understanding how communities and organizations perceive, plan for, and manage tourism development opportunities, how tourists learn from their travel experiences, and how to design more sustainable tourism experiences. She has published extensively on tourism and related areas with more than 200 refereed papers and book chapters. Her recent project areas include evaluating tourism as a tool for economic development in rural regions, tourist experience analysis, cross-cultural issues in tourism, designing effective tourist interpretation, and tourist storytelling. Her research is focused on understanding heritage interpretation in tourist settings, tourist storytelling, using quality of life as a way to investigate tourism impacts and community development, and new approaches to tourism planning and management at the destination level.

Lisa A. Pace is Senior Lecturer of Innovation Management and Foresight within the Edward de Bono Institute for Creative Thinking & Innovation at the University of Malta. After training in the natural sciences, Lisa completed a PhD in Innovation Management at the Manchester Institute of Innovation Research, Alliance Manchester Business School, UK. She has worked in the academic and public sectors on issues related to environmental innovation, energy efficiency, foresight, and stakeholder participation and has written in these areas.

Annarita Paiano holds a PhD degree in Commodity Science, Technology, Innovation, and Sustainable Production and is an associate professor in the Department of Economics, Management, and Business Law at the University of Bari Aldo Moro, Italy. She teaches courses on the technology of production cycles, as well as second-level economics and management. Her research is mainly concerned with sustainable tourism, bioenergy, energy policy, end-of-life management, life cycle assessment (LCA), and the circular economy.

Patricia Poehler holds a BSc degree in forestry from the University of Applied Forest Sciences in Rottenburg, Germany. Her thesis explores the importance of forest visits during the COVID-19 pandemic. She is working in forest administration in southern Germany.

Meenakshi Prasad currently holds the position of Assistant Professor at the University Department of Geography, Magadh University, Bodh Gaya. She is also a member of the DRC, Social Geography at A.N. Sinha Institute of Social Studies, Patna. She completed her B.A. (Hons.) and M.A. (Gold Medalist) at Patna University, Patna, and has received her PhD degree from the P.G. Department of Geography at Patna University. Her research interests include urban geography and environmental geography.

Juan Ignacio Pulido-Fernández received a Doctor Cum Laude degree in Economics from the University of Jaén and a Master's degree in Economic and Monetary Union from UNED. He completed his postgraduate training as a visiting researcher at the universities of Siena and Bologna (Italy), La Sorbonne (France), Leeds Metropolitan (England), and the National Autonomous University of Mexico. He is Associate Professor of Applied Economics (Tourism Economics) at the University of Jaén, where he also serves as Director of the Laboratory of Analysis and Innovation in Tourism (LAInnTUR) and coordinates the official Master's Degree in Sustainable Planning and Management of Tourism. He further serves as Director of the Chair of Inland Tourism at the University of Jaén. He has published more than 85 articles in journals that occupy relevant positions in lists such as SCI-SSCI and Scopus. His main research interests focus on tourism economics, destination management, the sustainability of tourism, tourism impacts, and social network analysis.

Vasco Santos holds a PhD in Business Sciences with a specialization in Marketing from Fernando Pessoa University, Portugal, a PhD in Tourism from Seville University, Spain, and an MSc in Marketing and Tourism Promotion from the Higher School of Tourism and Maritime Technology, Polytechnic of Leiria. He is Coordinating Professor of Tourism and Marketing at the Higher Institute of Management and Administration of Santarém (ISLA Santarém) as well as the coordinator of the degree in Tourism Management at the Higher Institute of Management and Administration of Santarém. In addition, he serves as a researcher in the Centre for Tourism Research, Development and Innovation (CiTUR) at the Polytechnic of Leiria. He has authored several papers and books on luxury tourism, sustainable tourism, ethical marketing, wine marketing, wine tourism, and employer branding.

Maria Sousa is a professor at the University Institute of Lisbon (ISCTE), a research fellow at the business research unit, and a collaborator of IPPS-ISCTE. She is an expert in digital learning and digital skills and held a post-doc position in this area from 2016 to 2018. She collaborates as an expert in digital skills with Deloitte (Brussels). She was a member of the coordinator committee of the PhD in Management at Universidade Europeia. She was also a senior researcher at GEE (Research Office) in the Portuguese Ministry of Economy and a competencies manager at AMA, IP, Public Reform Agency (Ministry of the Presidency and the Ministers Council). In addition, she served as a project manager at the Ministry of Labor and Employment and has participated in European projects of innovation transfer. She is an external expert for the COST Association—European Cooperation in Science and 3 Technology and former President of the ISO/TC 260.

Kyllikki Taipale-Erävala is an associate professor at the University of Eastern Finland. Her research focuses on the business survival of micro-businesses and SMEs in changing business environments. She is also interested in developmental aspects of micro-enterprises, innovation, business models, knowledge management, and advanced services. She has personal experience in the tourism industry, in the field of training and education at several school levels, the supervision of apprenticeships, as well as HPA and energy technology.

Deepika Chandra Verma is a fellow of the Indian Institute of Management Calcutta. She completed her BA (Hons) Economics and MA in Economics at Patna University (Patna). She was awarded the UGC-NET JRF scholarship in Economics in 2010 for the UGC-NET JRF exam held in December 2009.

Thomas Walker is a full professor of finance, the director and academic lead of the Emerging Risks Information Center (ERIC), the inaugural director of the Jacques Ménard/BMO Center for Capital Markets, and the Concordia University Research Chair in Emerging Risk Management (Tier 1) at Concordia University in Montreal, Canada. He previously served as an associate dean, department chair, and director of Concordia's David O'Brien Centre for Sustainable Enterprise. Prior to his academic career, he worked for firms such as Mercedes Benz, KPMG, and Utility Consultants International. He has written over 80 journal articles and books.

Janardan Krishna Yadav is an associate professor (marketing) at Jindal Global Business School. Before joining Jindal Global Business School, he was a visiting assistant professor at NIT Calicut where he taught customer relationship management. He received a university gold medal for his Master's in Tourism Administration (MTA) and was awarded a UGC-JRF award in Tourism Administration & Management. He is a fellow of IIM-Kozhikode with a specialization in marketing. He is a scientific committee member of the European Institute for Research and Development (EIRD).

List of Figures

List of Tables

Part I

Introduction

1

Sustainable Tourism: An Introduction

Thomas Walker, Ender Demir, Gabrielle Machnik-Kekesi, and Victoria Kelly

Introduction

In 2019, the tourism industry was among the fastest growing in the world (UNWTO, 2020). The United Nations' World Tourism Organization (UNWTO) recorded 1.5 billion international tourist arrivals globally—a 4% increase on the previous year—and predicted a similar increase for 2020 (UNWTO, 2020). In the same year, the industry not only accounted for 10.4% of the global GDP (US$9.2 trillion), but it was also responsible for significant job creation, with one in four new jobs across the world attributed to the tourism industry, accounting for 10.6% of all jobs (334 million) (World Travel and Tourism Council, 2022).

T. Walker (✉) • G. Machnik-Kekesi • V. Kelly
Concordia University, Montreal, QC, Canada
e-mail: thomas.walker@concordia.ca; gabrielle.machnik-kekesi@concordia.ca; victoria.kelly@concordia.ca

E. Demir
Reykjavík University, Reykjavik, Iceland
e-mail: enderd@ru.is

T. Walker et al. (eds.), *Sustainable Tourism*,
https://doi.org/10.1007/978-3-031-43528-7_1

However, in 2020, the modern tourism industry faced unprecedented challenges due to COVID-19. The pandemic, which forced the closure of many international borders and brought tourism activities to a halt, had devastating consequences for the industry's growth. Though the UNWTO (2022) reported a 4% increase in international arrivals in 2021, this figure still represented a 72% drop from pre-pandemic times. As 2022 drew to a close, the tourism industry started to rebound, reaching nearly 60% of its pre-pandemic levels of activity (UNWTO, 2022); by the end of 2023, international tourism figures were at 88% of their pre-pandmic levels (UNWTO, 2024). During this transitional recovery period, conversations surrounding sustainable tourism practices have become all the more urgent.

As tourism continues to recover from the severe economic impacts of COVID-19, questions arise as to how this renewed sense of urgency in the industry can be mobilized to the benefit of sustainability. While the tourism industry provides economic and social benefits and has the potential to contribute to sustainable development through job creation and economic redistribution, it also has many problematic side effects. Conventional tourism can encroach on habitats, ecosystems, and wildlife, as well as accelerate extractivist forms of land use and development (Davies & Cahill, 2000).

Experts have voiced concerns as to how the negative effects of such tourism practices could be mitigated as early as 1973 (Berno & Bricker, 2001). The concept of sustainable tourism emerged in 1987 at the World Commission on Environment and Development, where ideas of sustainable development and its implications in tourism first began to burgeon (Berno & Bricker, 2001). The United Nations Environment Programme and World Trade Organization (2005) describe sustainable tourism as making optimal use of environmental resources, maintaining essential processes, and helping conserve natural heritage and biodiversity. Moreover, it respects the socio-cultural heritage of local communities, contributes to inter-cultural understanding and tolerance, and ensures long-term economic operations by providing fairly distributed economic benefits among all participants. The promotion and development of effective sustainable tourism measures not only limit the negative impacts associated with tourist travel, but also inspire ways to conserve and perhaps even rebuild this important industry in a way that respects cultural difference, economic justice, and environmental preservation. This edited collection collates explorations of both new and long-standing challenges associated with sustainable tourism and aims to reconcile how policies and frameworks can establish a strong foundation for the future of this industry. The tourism industry is multifaceted, as reflected by the interests of its many stakeholders and as observed through the lenses of various disciplines. This book highlights this diversity by presenting chapters which examine tourism's various impacts not only from an economic perspective, but also from an innovation-driven and management-based perspective.

Overview of Content

The above section briefly introduces the current challenges faced by the tourism industry and highlights that the sector faced many hurdles with regard to sustainability long before the COVID-19 pandemic. The pandemic, it can be argued, merely served as a catalyst for the industry's need to address these issues; some of the chapters in this book explore the impacts of this seismic event directly, while others address different, but equally pressing, topics. The book is organized as follows.

Part I: Sustainable Tourism Frameworks

The first section of this collection explores sustainable tourism frameworks and aims to contextualize sustainability in the ever-growing tourism industry. These chapters are linked in their capacity to propose models and frameworks that consider sustainable tourism specifically with respect to destination.

The section begins with Chap. 2, *Sustainability-Oriented Innovation in Tourism Destinations: Born Ready or Bred for Success?* in which author **Lisa Pace** situates sustainability-oriented innovation within the domain of tourist destinations. Focusing on four core capabilities—leadership, collaboration, marketing capacities, and capabilities for sustainable innovation management—Pace argues that tourist destinations stand to benefit from strategic development in creating sustainability-oriented innovations that increase the appeal of tourism products.

Chapter 3, *Sustainable Livelihoods and Tourism Businesses: An Overview of the Nexus in Developing Countries*, suggests operational and institutional indicators as well as variables that can be used to understand the relationship between sustainable tourism and sustainable livelihoods on a global scale. Author **Annarita Paiano**'s replicable model comprises a valuable tool for tourism sector stakeholders.

Finally, Chap. 4, *Innovations in Luxury Tourism: Developing a Novel Theoretical Framework* offers a mixed-methods study by authors **Vasco Santos**, **Carlos Costa**, **Manuel Au-Yong**, and **Maria Sousa.** They develop an original model that aims to incorporate sustainability into a luxury tourism paradigm for the use of both researchers and industry professionals.

Part II: Sustainable Tourism Innovations and Solutions

Focusing on technological advances as well as the reinvention of tourism activities, this section explores new and innovative ways to achieve sustainability; local tourism is an undercurrent throughout the section.

Chapter 5, *Recreation in Forests: Implications from the COVID-19 Pandemic*, showcases the importance of recreational tourism, which became painfully apparent during the course of the pandemic. Authors **Patricia Poehler** and **Monika Bachinger** examine visits to forests in Germany during the COVID-19 pandemic to explore the possibility that domestic travel to green spaces can enhance well-being. Based on their findings, the authors propose relevant destination management strategies.

In Chap. 6, *Staycations: A Sustainable Tourism Solution?*, **Gianna Moscardo** adopts a system thinking approach to conceptualizing "staycationing" as an alternative to recreation and tourism. Moscardo proposes decision-making strategies that can encourage staycations and thus more sustainable tourism practices.

Continuing the thread of local tourism, Chap. 7, *Local Tourist Shopping: How Purchases in a Destination Can Contribute to Sustainability*, uncovers how shopping experiences both contribute to the appeal of various tourist destinations and have the potential to exacerbate tourism's unsustainability. Authors **Alba García-Milon** and **Emma Juaneda-Ayensa** propose local tourist shopping as a concept that can enhance tourism's overall sustainability and make recommendations for fostering this practice.

Further building on this section's focus on locality, Chap. 8, *The Role of Smart and Experimental Tourism in Sustainable Development*, evaluates Portugal's preparedness to engage with a rapidly changing tourist sector in the wake of COVID-19 and in the midst of technological change. Author **Silvia Fernandes** argues that through smart city acceleration programs, the Internet of Things (IoT), and augmented reality, the country can increase its resilience and encourage sustainable practices.

Finally, in Chap. 9, *A New Tool for Smart Tourism Destinations: Indicators of Sustainable Intelligence*, **Yaiza López-Sánchez**, **Juan Ignacio Pulido-Fernández**, and **José Luis Durán-Román** nuance the relationship between smart tourist destinations (SDs) and sustainable tourism indicators through their own model that incorporates tourists themselves in SD sustainability. The authors base their "Indicator of Tourists' Sustainable Intelligence" on a two-round Delphi survey and provide useful information on the sustainable management of SDs.

Part III: Sustainable Tourism Policies and Practices

This final section of the book analyzes existing global policies and practices in place regarding sustainability in tourism activities. The chapters determine the efficiency of these practices and offer suggestions with regard to global policy making.

The section begins with Chap. 10, *Tourism Views and Experiences of Micro- and Small Enterprises in Northern Finland and Russian Karelia during the COVID-19 Pandemic: A Cross-Border Operations Perspective*, in which authors **Katariina Ala-Rämi** and **Kyllikki Taipale-Erävala** share the experiences of Finnish and Russian micro-tourism entrepreneurs during the pandemic and before the invasion of Ukraine, showcasing both challenges and opportunities of cross-EU border tourism. The authors also recommend ways to increase economic activity and rural revitalization.

In Chap. 11, *Environmental and Tourism Policies for the Decarbonization of the Tourism Sector*, **Yoshihiro Hamaguchi** suggests that, in spite of sustainable tourism's increased traction, the tourism sector remains carbon-intensive. Through a literature review of the topic, Hamaguchi uncovers controversies surrounding the tourism-led growth hypothesis (TLGH) and proposes environmental and tourism policies that could activate the decarbonization of the tourism sector, particularly in the context of increased travel at this stage of post-COVID-19 pandemic recovery.

In chap. 12, *Can Tourist Taxes Make Tourism More Sustainable?*, authors **Juan Ignacio Pulido-Fernández**, **José Luis Durán-Román**, and **Yaiza López-Sánchez** employ a case study of an Andalusian tourist business transitioning to sustainability to explore which factors and variables impact customers' "willingness to pay." The authors' findings contribute to our understanding of sustainable destination management and tourist experience management.

Finally, Chap. 13 concludes this section by analyzing policies at the nexus of walkability, gender, spirituality, and travel in Bodh Gaya, India. In *Walking in Bodh Gaya: A Gendered Approach to the Analysis of Sustainable Tourism Policies*, authors **Deepika Chandra Verma**, **Janardan Krishna Yadav**, and **Meenakshi Prasad** build on findings from a focus group and offer suggestions with respect to urban planning and the management of local/tourist dynamics.

While dedicated to various topics, there are noticeable trends and patterns across chapters in different sections. In addition to the pandemic serving as a throughline for multiple studies and as the main focus in Chap. 10, place-specific models and solutions, as represented by Chaps. 5 and 12 (among others) provide valuable insights, especially when considered alongside more macro-scaled studies, exemplified by contributions including Chaps. 3 and 11. Building on concepts of place, Chap. 9's focus on destination can effectively be contemplated alongside Chap. 6's study of staycationing. In addition, the central role played by management and governance, as Chaps. 2 and 4 illustrate, is a recurring theme in the collection. The cultural and affective appeal of tourist sites is sensitively presented in Chaps. 7, 8, and 13, the latter introducing the key concern of how gender intersects with sustainable travel considerations. With broad geographic representation and a rich variety of thematic explorations, this collection of essays shares fresh findings while providing the basis for further research on the vitally important topic of how to experience the wonders of travel while doing so sustainably.

References

Berno, T., & Bricker, K. (2001). Sustainable tourism development: The long road from theory to practice. *International Journal of Economic Development, 3*(3), 1–18.

Davies, T., & Cahill, S. (2000). *Environmental implications of the tourism industry*. Retrieved May 12, 2024, from https://www.csu.edu/cerc/documents/EnvironmentalImplicationsTourismIndustry2000.pdf

United Nations Environment Programme, & World Trade Organization (WTO). (2005). *Making tourism more sustainable: A guide for policy makers*. https://wedocs.unep.org/20.500.11822/8741

United Nations World Tourism Organization (UNWTO). (2020). *International tourism growth continues to outpace the global economy*. January 20. Retrieved November 6, 2022, from https://www.unwto.org/international-tourism-growth-continues-to-outpace-the-economy

United Nations World Tourism Organization (UNWTO). (2022). T*ourism grows 4% in 2021 but remains far below pre-pandemic levels*. Retrieved November 6, 2022, from https://www.unwto.org/news/tourism-grows-4-in-2021-but-remains-far-below-pre-pandemic-levels

United Nations World Tourism Organization (UNWTO). (2024). https://www.unwto.org/news/international-tourism-to-reach-pre-pandemic-levels-in-2024#:~:text=Following%20a%20strong%202023%2C%20international,estimated%201.3%20billion%20international%20arrivals

World Travel & Tourism Council. (2022). *Travel & tourism economic impact: World travel & tourism council*. Retrieved November 6, 2022, from https://wttc.org/research/economic-impact#:~:text=Meanwhile%2C%20international%20visitor%20spending%20amounted,21.7%25%20rise)%20in%202021

Part II

Sustainable Tourism Frameworks

2

Sustainability-Oriented Innovation in Tourism Destinations: Born Ready or Bred for Success?

Lisa A. Pace

Introduction

Tourism is at a crossroads: tourism operators can either continue on a path of mass consumption with consequences such as the deterioration of natural and social capital or, they can "take responsibility for the long-term sustainability of ecosystems by adopting green strategies even if they increase the cost of services" (Camisón, 2020, p. 1). This statement reflects an underlying tension of sustaining the growth of tourism whilst alleviating its impacts on natural and socio-cultural heritage and on quality of life. It is beyond the scope of this chapter to synthesize the arguments framing the integration of sustainability principles in tourism. Instead, the aim is to develop a perspective on how destinations re-orient along a sustainable path by developing capabilities to innovate for this purpose.

L. A. Pace (✉)
The Edward de Bono Institute for Creative Thinking and Innovation, University of Malta, Msida, Malta
e-mail: lisa.a.pace@um.edu.mt

T. Walker et al. (eds.), *Sustainable Tourism*,
https://doi.org/10.1007/978-3-031-43528-7_2

At the outset, sustainable development in tourism has been mainly concerned with the conservation and preservation of resources. Sharpley (2020) claims this to be a response to the negative connotations of tourism activities—especially mass tourism—on destination attractiveness. This has led to the development of alternative forms of tourism such as eco, green, and slow tourism. However, these generally remain niche areas, attracting a cohort of "green" or "environmentally aware" customers; compared to "mainstream" tourism, they do not have a significant impact on the competitiveness of a destination in the long term (Saarinen, 2006; Sharpley, 2020). Perles-Ribes and Ivars-Baidal (2018) outline a series of general principles underlying the integration of sustainability in tourism destinations: the sustainable management of the natural environment is at the top of the list. Next is the role of sustainability in contributing to the well-being of residents and tourists alike and the significance for destinations of understanding the market and the demand for tourism experiences that respect the characteristics of the destination. Aspects linked to tourism's contribution to local economies and innovations that impact on the well-being of local communities are considered integral to discourse on sustainable development in tourism (Buhalis, 2000; Crouch & Ritchie, 1999; Cucculelli & Goffi, 2016).

Innovation for sustainability has gathered considerable interest in tourism research, particularly in relation to investigating the drivers of environmental innovation and exploring eco-innovative practices in tourism (Pikkemaat et al., 2019). Environmental or "eco-innovation" is largely concerned with the development of new or improved products, services, and management processes that reduce or avoid harmful environmental impacts (Kemp & Pearson, 2007; Rennings, 2000). These innovation activities accrue positive economic returns to industries investing in clean tech processes and resource efficient practices. The social pillar of sustainability has recently attracted the interest of researchers and practitioners exploring the contribution of social enterprises and social entrepreneurs to circularity principles and to the co-creation of social innovations that have a positive impact on local communities (Pikkemaat et al., 2019). In this chapter, the term "sustainability-oriented innovation" is utilized to acknowledge the normative dimension of innovation linked to achieving social goals and the well-being of societies and populations. Sustainability-oriented innovation also has a temporal dimension in looking towards future generations and meeting their resource needs (Godin & Gaglio, 2019). Indeed, sustainability-oriented innovation is often regarded as a solution to grand social and environmental challenges and to re-orienting growth in economies and societies along sustainable pathways (Godin & Gaglio, 2019).

Tourism destinations gain a competitive advantage from strategically deploying competencies and resources to develop sustainability-oriented innovations that enhance the attractiveness of the tourism product (Dwyer & Kim, 2003; Paunovic et al., 2020). This involves exploiting the natural resource base and socio-cultural capital to create innovative experiences that are both economically profitable and have a positive impact on environmental and social dimensions. This chapter characterizes sustainability-oriented innovation in the context of tourism destinations and proposes a framework of four core capabilities—these are leadership, collaboration, marketing capacity, and sustainable innovation management capabilities—that together create an enabling environment for tourism destinations to innovate for sustainability. Drawing from literature on strategic management and innovation, the chapter assesses how destinations adapt and respond to the challenge of sustainable development and offers a rationale for tourism destinations to mobilize assets and capabilities for sustainability-oriented innovations. It concludes with suggestions for policymakers and tourism practitioners on how tourism destinations can achieve a transition towards sustainable development.

Characterizing Sustainability-Oriented Innovation in Tourism

When discussing innovation in tourism, it is relevant to recall the characteristics of tourism services as being intangible and often co-produced through encounters and interactions with different tourism suppliers (Buhalis, 2000; Crouch, 2011; Evans, 2016; UNWTO, 2019). Innovation in tourism may arise as a new service concept, be it a new product or service. It can come about through establishing new modes of interacting with the customer and delivering the service, such as through the use of information communication technology (ICT), or establishing relations with new business partners or through marketing innovations (Font & McCabe, 2017; Weidenfeld, 2018).

Sustainability-oriented innovations in the tourism industry have developed principally in response to regulatory pressures, e.g., related to planning, pollution control, heritage conservation, and efficient resource consumption (Buckley, 2012). Green or environmental technologies and environmental management systems are deployed to enhance operational efficiency by implementing energy efficient and water conservation measures. Voluntary certification and labelling schemes and community outreach programmes are implemented amongst tourism firms as a marketing strategy and to measure environmental performance, particularly in accommodation establishments (Gonçalves de Almeida et al., 2021). New opportunities for local businesses are established, e.g., by promoting traditional gastronomy and local produce through partnerships established between local producers, artisans, and restaurants.

In 2018, the European Commission launched an EU-wide competition for European cities to showcase innovations in tourism in areas relating to digitalization, accessibility, sustainability, and the protection of cultural heritage. Over the years, several cities have been awarded the "European Capital of Smart Tourism".[1] The competition highlights the broad variety of initiatives and innovations implemented to achieve sustainability in tourism. Several of these initiatives highlight a growing interest in exploring the applications of digital technologies to improve access to information on cultural heritage and enhance the accessibility of cultural attractions, including events and festivals. ICTs and disruptive technologies—such as big data and artificial intelligence—are being deployed to develop innovative services for managing tourism flows in city centres and leisure facilities. One example is the "virtual queuing system" implemented in Gothenburg to monitor the flow of visitors and control their access to popular cultural sites. Virtual reality enables new

[1] Information on 'The European capital of Smart Tourism' was accessed here on 20 May 2022: https://smart-tourism-capital.ec.europa.eu/index_en

ways of experiencing city attractions such as using 3D city maps that offer enhanced visual and audio experiences of city attractions (EC, 2020). Digital hubs and clusters provide a space for entrepreneurs to develop innovative digital applications in tourism, and initiatives such as the hackathons organized in Copenhagen and the "Museomix" hackathon in Brussels have gained interest among professionals and amateurs engaged in envisioning innovative experiences for the future of tourism (EC, 2020).

Cities are upgrading traditional cultural attractions by incorporating sustainability elements, such as the "kayak and paddle boarding" experience in Lyon offering an alternative way to discover the city's natural heritage, or saunas in Helsinki running on alternative energy sources (EC, 2020). Tourism activities may indirectly contribute to improving quality of life for citizens by preserving biodiversity and heritage sites (Di Bella et al., 2019; Foronda-Robles et al., 2020; González-Reverté, 2019; Perles-Ribes & Ivars-Baidal, 2018). Some initiatives aim at repurposing public buildings as cultural sites such as the Centre for Contemporary Arts in Malaga that replaces the former wholesale market.

The examples illustrate that sustainability-oriented innovations include broad categories of innovation types that range from tangible products and attractions to novel management practices, service offerings, and experiences that together contribute to shaping the attractiveness of the destination. These have been studied at the firm and destination levels, emphasizing different aspects related to operational eco-efficiency, the conservation of socio-cultural heritage, customer engagement, and pro-environment behaviors, as well as exploring the benefits of "eco" or sustainability innovation on local communities and their well-being. Firms may innovate by making incremental improvements based on available knowledge, such as in the case of eco-efficiency. Alternatively, novel tourism products and experiences may be implemented that require firms and destinations to rethink their strategies and business models.

Building Capabilities for Sustainability-Oriented Innovation in Destinations

How do tourism destinations innovate for sustainability? A destination is a physical space where visitors live and consume an experience (Seguí-Amortegui et al., 2019; UNWTO, 2019). It is delineated by the infrastructures, services, and facilities necessary to deliver the tourism product (Bornhorst et al., 2009; Pechlaner et al., 2009). Reinhold et al. (2015) are somewhat critical of this definition as being too supply-oriented and focussed on "what is offered where and by whom" (p. 138) and argue that it neglects the role of the tourist and the demand side in developing tourism experiences. Increasingly, destinations are conceptualized as a space where a diversity of tourism actors is involved in co-producing tourism products and services (Reinhold et al., 2015).

Destinations accrue a strategic advantage from their physical location and natural assets—the latter refer principally to the destination's climate, culture, history, and other core attributes such as infrastructures and accessibility that together determine its attractiveness (Camisón & Fores, 2015; Crouch & Ritchie, 1999). Based on an expert survey administered in Gran Canarias, Rodríguez-Díaz and Espino-Rodríguez (2016) identify three key factors that influence a destination's performance. These are "key resources and supply chain" (the climate, beaches and the sea, airports, tour operators), complementary leisure facilities (including sports, shopping centres, and cultural attractions), and governance and security. However, relying on core resources alone is unlikely to sustain the economic growth of a destination in the longer term. Destinations are compelled to innovate by identifying unique opportunities to exploit endogenous resources and assets and create experiences that are superior to those in competing destinations to gain a competitive advantage (Camisón et al., 2016; Dwyer & Kim, 2003; Paunovic et al., 2020; Weidenfeld, 2018). These resources may include tangible assets such as geographic location, infrastructures and attractions, and accommodation, as well as intangible assets that include human resources, expertise and know-how, as well as organizational assets such as brand image and reputation (Eisenhardt & Martin, 2000).

Tourism stakeholders have different views about how best to exploit destination resources and preserve these for future generations. An additional consideration is that of ensuring a fair or equitable distribution of benefits from tourism activities amongst all relevant stakeholders (Buhalis, 2000). The decision about which core attributes of a destination to promote is determined by destination management organizations (DMOs) and influenced by the strategic choices made by diverse actors operating in the tourism industry. To this end, destinations need to create new and value-added services and products. Indeed, innovating for sustainability is a way to achieve this. Sustainable development is often considered central for destinations to diversify the tourism product and create new and improved tourism experiences, with the goal to enhance competitive advantage in the long term (Crouch, 2011; Paunovic et al., 2020; Seguí-Amortegui et al., 2019).

In practice, a destination's capability to mobilize resources for sustainability-oriented innovation depends on a combination of factors. The stage of destinations' lifecycle largely determines whether new products are developed or whether the destination enters new market segments. Saarinen (2006) further elaborates on this and states that a destination's resource needs and intensity of resource use change with different stages of the tourism lifecycle; understanding these changes may be key to achieving resource-based sustainability. Also, the destination's market orientation and whether this promotes high-value tourism products or mass tourism consumption is another factor (Buhalis, 2000; Romão, 2020). Notwithstanding, destinations need to recognize where and how to creatively recombine resources towards innovative value-creating activities and products to contribute to sustainable development (Sainaghi et al., 2019).

Destinations need to determine which resources to mobilize, when, and how to activate these to achieve sustainability outcomes. Resources and competencies are often dispersed among actors at different levels. These are the (eco)system level, the cluster or district level, and the firm level. Integrating resources across these levels needs to be part of a coordinated strategy (Da Cunha Bezerra et al., 2020). The following section explores the role of a destination's capabilities in the development of sustainability-oriented innovations and their contribution towards enhancing the tourism product. These capabilities enable destinations to deploy resources and recombine these in innovative ways to improve or develop new tourism services and experiences.

A "Capabilities" Framework for Sustainability-Oriented Innovation in Destinations

This chapter proposes a framework of core capabilities that enable a destination to implement sustainability-oriented innovations and diversify the tourism product, per Fig. 2.1. It is based on a review of the literature on tourism, sustainability, and competitiveness as well as on addressing innovation and dynamic capabilities. The elements of the framework are discussed below.

Destination Leadership

Leadership is critical in steering a destination towards a sustainable path. Beritelli and Bieger (2014) describe destination leadership capability as a systemic property. Rather than being rooted in a particular entity,

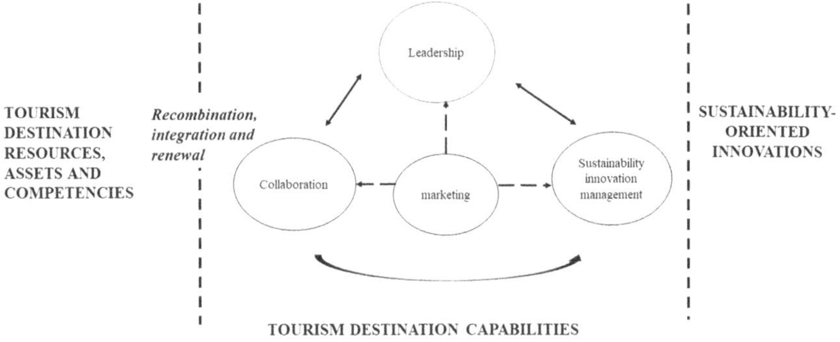

Fig. 2.1 Framework linking a tourism destination's capabilities for sustainability-oriented innovation

leadership is an inter-relational property of actors "mutually influencing one another" to achieve a common objective (Beritelli & Bieger, 2014). Due to their complex structure, destinations are often characterized by dispersed initiatives of tourism entities operating autonomously and lacking a strategic orientation (Camisón et al., 2016). Leadership capability contributes to developing a collective vision on a destination's strategic position in the market and its sustainability priorities. A vision with clearly defined objectives for sustainability spurs stakeholders to take action and prioritize initiatives that create new value for the tourism experience (Graci, 2013).

In an investigation on the adoption of sustainability-oriented innovations, Inigo and Albareda (2019) demonstrate that firms with leadership capabilities are better able to connect with societal and environmental stakeholders outside their industrial sector and in so doing contribute to building ecosystems oriented towards sustainability innovation. In a similar vein, a destination's leadership capability mobilizes a collective of tourism and non-tourism actors to achieve common objectives for sustainable development (Beritelli & Bieger, 2014). The development of a leadership capability leads to a collective vision on the goals and on the direction of development—fulfilling a path steering function suggested by Baggio et al. (2010). Beritelli and Bieger (2014) describe the role of leadership as bringing together tourism actors with different interests to agree to a "common ground" (p. 33) when responding to challenges affecting a destination's development. In this context, leadership capability may facilitate the access to and allocation of resources for the purpose of innovation by spearheading collaborative initiatives on the development of innovative tourism services and sustainability initiatives (Buhalis, 2000; Zehrer et al., 2013).

Collaboration and Sustainability-Oriented Innovation

Collaboration among tourism actors facilitates "creative combinations of unique and inimitable resources" (Denicolai et al., 2010, p. 262) that firms would not be able to access alone. Graci (2013) describes collaborations as loosely coupled partnerships between tourism organizations and individuals sharing a common goal—that of identifying innovative tourism solutions for sustainable development. Steering a destination towards a sustainable path may require new governance mechanisms involving different structural arrangements of tourism actors, and new relations between stakeholders may be important to consider that shape the actions and initiatives adopted by tourism actors (Baggio et al., 2010; Zollo et al., 2013). For example, Gill and Williams (2011) discuss the role of innovative governance models in strategic resort management and demonstrate the benefit of stakeholder engagement in incorporating societal issues and community concerns in the sustainable development and planning of the destination.

Gonçalves de Almeida et al. (2021) refer to collaboration as a means for tourism organizations working together to achieve mutual benefits. The latter may relate to new investments, knowledge-sharing, or building complementary resources (Sainaghi et al., 2019). Collaborative capabilities describe an ability to mobilize assets and resources within networks or partnerships to facilitate the flow of information and knowledge about innovation (Choi & Hwang, 2015; Dyer & Singh, 1998). Firms that build collaborative partnerships have better access to environmentally relevant knowledge and know-how about how to integrate sustainability into new product design or for greening industrial processes (Adams et al., 2016; Dangelico et al., 2013; Pace, 2016). Likewise, Camisón et al. (2017) delineate collaborative capabilities in tourism clusters or districts as facilitating access to new ideas and information needed to generate innovations and to broaden the range of products and services offered by the destination.

Collaborative capabilities may strengthen complementary knowledge and skills by promoting the diffusion and exploitation of existing knowledge that may lead to incremental innovation and the adoption of best practices (Camisón et al., 2017; Gutiérrez-Rodríguez et al., 2020). Alternatively, these capabilities renew the knowledge base by enabling new skills and competencies to develop within tourism clusters or destinations (Gonçalves de Almeida et al., 2021; Gutiérrez-Rodríguez et al., 2020). Firms operating in a tourism cluster or district have been shown to benefit from the knowledge and experience shared among members in the cluster. However, tourism firms need to develop enhanced absorptive capacities to assimilate knowledge about sustainability and innovation within their routine operations (Camisón et al., 2016; Gutiérrez-Rodríguez et al., 2020; Pace & Miles, 2020).

Marketing and Branding

Marketing and branding are considered central to the attractiveness of a destination and to establishing an image that is appealing to the visitor (Buhalis, 2000; Rodríguez-Díaz & Espino-Rodríguez, 2016). Marketing strategies are designed to highlight the "uniqueness" of a destination and its core resources that include its natural, social, and economic assets (Crouch, 2011; Dwyer & Kim, 2003). Although often dismissed as greenwashing or as colluding with the short-termism and profit-oriented imperatives that come with tourism marketing strategies, sustainability marketing has an important role to play in creating and promoting "more sustainable tourism offerings" (Font & McCabe, 2017, p. 871).

One of the aims of sustainability marketing is to communicate a destination's eco or sustainability profile and integrate this into the destination's brand image (Boes et al., 2016; Haugland et al., 2011; Lemmetyinen & Go, 2009). Further to this, adopting a sustainability perspective in tourism marketing involves promoting a shift from mass consumption and resource use towards encouraging "preservation and pro-environmental behaviours" (Jamrozy, 2007, p. 125). Indeed, Font and McCabe (2017) discuss how destinations need to re-orient their marketing and communication resources to potentially influence sustainable consumption behaviours or to enhance the attractiveness of sustainability products amongst customers. Furthermore, in an analysis of environmental marketing strategies in the Greek hotel sector, Leonidou et al. (2013) argue for the importance of hotels' developing a shared marketing strategy—one that communicates a common set of goals and objectives—about sustainability. This spurs businesses and other tourism actors to strategically reorganize their resources, competencies, and services to achieve similar goals for sustainability in the destination.

Guests are taking on a more active role in co-creating tourism experiences. Destinations require capabilities to better understand customer needs and expectations and how these factors might impact the demand for and pressure on a destination's core assets and resources (Cucculelli & Goffi, 2016; Hassan, 2000; Saarinen, 2006). Sensing capabilities on new customer trends and "eco-preferences" enable a timely response to changes in the market. Hence, destinations build capacity to adapt and transform their tourism products towards sustainability (Camisón et al., 2016; Leonidou et al., 2013). For example, Rodríguez-Díaz and Espino-Rodríguez (2016) demonstrate that in the sun and sea destination of the Gran Canarias (Spain), customers tend to prioritize natural resources and complementary attractions, including climate, sea, beaches, and tour operator services, over cultural resources, whereas destinations with a higher natural resource capital are likely to attribute higher importance to societal and cultural assets. This requires tourism operators to develop a better understanding of customer preferences and to promote those sustainability attributes and experiences that differentiate a destination from its competitors (Jamrozy, 2007). Sustainability marketing capabilities may create opportunities for developing new product offerings. However, as discussed earlier, there needs to be a common understanding of which elements of sustainability to prioritize and whether these relate to social equity, economic viability, or environmental protection (Kemper & Ballantine, 2019).

Sustainability-Oriented Innovation Management

Innovation management capabilities enable destinations to develop and deliver sustainability-oriented innovations that are part of the portfolio of services and attractions constituting the tourism experience. Prats et al. (2008) maintain that innovation is central to the development of a tourism destination comparing this to a local innovation system. Using this analogy, a tourism destination is seen to function as a "system" of interconnected actors that together create new value for the destination by enabling the flow of knowledge about innovative products and services (Boes et al., 2016; Camisón & Fores, 2015). Innovations arise when actors collaborate in the co-development of innovative tourism products, typically enabled by the use and applications of ICTs (Boes et al., 2016; Rodríguez-Díaz & Espino-Rodríguez, 2008).

Innovation in tourism destinations has been described as open and distributed. This in part reflects the fact that destinations consist of a multiplicity of actors directly linked to the tourism industry and those operating in complementary industries such as entertainment and leisure, food service supply, transportation, etc., that are involved in developing and delivering tourism products (Hassan, 2000; Haugland et al., 2011). Tourism firms can take advantage of the knowledge, resources, and capabilities available within the tourism destination by establishing collaborations with non-business actors including universities, research organizations, and non-governmental organizations. Collaboration may take different forms ranging from sharing information about needs and expectations, cooperating in order to manage tensions and achieve preset goals, through to intensive collaborative partnerships wherein organizations exhibit high interdependent skills and capabilities (Choi & Hwang, 2015; Dyer & Singh, 1998). Public institutions and other local agents (e.g., tourism boards) act as intermediaries in stimulating or strengthening collaboration among firms when implementing innovative technologies, management, and organizational practices (Camisón & Fores, 2015).

To innovate, firms require access to markets and technologies as well as to the necessary resources such as skills and a research capacity to enable innovations to develop and diffuse. Another aspect relates to the availability of support infrastructures, including access to finance and business advisory services supporting the development of sustainability products. Tourism firms are generally not well-embedded in an innovation system infrastructure (Hjalager, 2010). They implement technologies developed elsewhere and adapt these to create new or improved tourism experiences. Indeed, knowledge about environmental technologies and sustainability practices is not always at the core of tourism firms' activities so that these often rely on external knowledge sources to implement sustainability initiatives, accessible through the supply chain or through informal networks (Adams et al., 2016; Dangelico et al., 2013; Pace, 2016; Pace & Miles, 2020). To this end, tourism firms need to develop collaborative capabilities within supply chains in order to bridge knowledge gaps and exploit business links and networks to address environmental challenges and co-produce environmental solutions (Paulraj, 2011).

The framework provides a basis for characterizing how sustainability-oriented innovations arise in tourism destinations. It aims to address the question of why some destinations develop these innovations, whereas others might not. The adoption of sustainability-oriented innovations cannot be explained merely by the presence of core resources and assets—these are environmental, social, and cultural resources. It is contingent on the type and extent of capabilities and competencies redeployed by destinations to successfully create these innovations. This is not to say that sustainability-oriented innovation is a result of the presence (or absence) of a set of capabilities. As discussed earlier, the characteristics of a destination, its lifecycle, and market orientation influence how resources are redistributed for innovation. Destinations need to balance the use of their capabilities and resources based on fluctuations in market demand and economic perturbations as well as on the response to exogenous shocks.

Conclusion

There are multiple opportunities for tourism destinations to develop sustainability-oriented innovations and enhance the attractiveness of the tourism experience. This chapter develops a framework for linking a destination's capabilities to adopting these innovations. Leadership capabilities spur destination actors to implement a common vision for the future sustainable development of the territory and to develop a strategy that involves marketing the destination's natural and cultural resources and strategic attributes. The framework suggests that destinations are able to identify unique opportunities to implement innovations for sustainability by developing capabilities to establish collaborative partnerships within the tourism supply chain and capabilities to tap into innovative markets, technologies, and know-how.

There are several related areas that could benefit from further research. One is the relationship between destination capabilities and the type of sustainable innovations adopted. Destinations may leverage different capabilities based on their characteristics and assets, and specific capabilities may be more relevant to implement radically new tourism experiences over making incremental improvements. A research opportunity lies in testing the capabilities framework in destinations at different stages of their lifecycle, such as in emerging and mature destinations and those with different market orientations.

Responsibility for the development of sustainable tourism experiences cannot fall on destination management organizations (DMOs) alone. In the future, many more stakeholders operating in destinations are likely to play a leading role in co-creating tourism experiences, including for sustainability-oriented innovation. The successful development and management of innovation hinges on establishing collaborations and partnerships between municipalities and city councils, tourism businesses and associations, and civil society organizations to synergize efforts when pursuing innovative ideas to solve tourism-related challenges. Innovative clusters are seen as an opportunity for entrepreneurs and start-ups working in diverse sectors including digitalization, sustainability, and the circular economy to establish synergies with tourism businesses for the co-creation of novel tourism experiences. DMOs could act as enablers and facilitators in co-producing these experiences and in managing new or different configurations of actors (Reinhold et al., 2015). There are obstacles to achieving innovation for sustainability in tourism. Tourism managers need to draw from experiences and lessons learnt from the successful implementation of service innovations more generally and learn how to develop concepts of sustainability-oriented innovation that are transferrable to the tourism context and to upscale these initiatives to the destination level. When planning tourism policy, decision-makers need to have a clear vision on the role of sustainability for a destination's growth. This should lead to developing a potential pipeline of innovations and novel tourism experiences and a roadmap to achieve these.

References

Adams, A., Jeanrenaud, S., Bessant, B., Denyer, D., & Overy, P. (2016). Sustainability-oriented innovation. *A Systematic Review International Journal of Management Reviews, 18*, 180–205.

Baggio, R., Scott, N., & Cooper, C. (2010). Improving tourism destination governance: A complexity science approach. *Tourism Review, 65*(4), 51–60.

Beritelli, P., & Bieger, T. (2014). From destination governance to destination leadership—Defining and exploring the significance with the help of a systemic perspective. *Tourism Review, 69*(1), 25–46.

Boes, K., Buhalis, D., & Inversini, A. (2016). Smart tourism destinations: Ecosystems for tourism destination competitiveness. *International Journal of Tourism Cities, 2*(2), 108–124.

Bornhorst, T., Ritchie, B. J., & Sheelan, L. (2009). Determinants of tourism success for DMOs & destinations: An empirical examination of stakeholders' perspectives. *Tourism Management, 31*(5), 572–589.

Buckley, R. (2012). Sustainable tourism: Research and reality. *Annals of Tourism Research, 39*(2), 528–546.

Buhalis, D. (2000). Marketing the competitive destination of the future. *Tourism Management, 21*, 97–116.

Camisón, C. (2020). Competitiveness and sustainability in tourist firms and destinations. *Sustainability, 12*, 2388.

Camisón, C., & Fores, B. (2015). Is tourism firm competitiveness driven by different internal or external specific factors?: New empirical evidence from Spain. *Tourism Management, 48*, 477–499.

Camisón, C., Forés, B., & Boronat-Navarro, M. (2017). Cluster and firm-specific antecedents of organizational innovation. *Current Issues in Tourism, 20*(6), 617–646.

Camisón, C., Puig-denia, A., Forés, B., Fabr, M. E., Muñoz, A., & Muñoz-Martínez, C. (2016). The importance of internal resources and capabilities and destination resources to explain firm competitive position in the Spanish tourism industry. *International Journal of Tourism Research, 18*, 341–356.

Choi, D., & Hwang, T. (2015). The impact of green supply chain management practices on firm performance: The role of collaborative capability. *Operations Management Research, 8*, 69–83.

Crouch, G. I. (2011). Destination competitiveness: An analysis of determinant attributes. *Journal of Travel Research, 50*(1), 27–45.

Crouch, G. I., & Ritchie, B. J. R. (1999). Tourism, competitiveness, and societal prosperity. *Journal of Business Research, 44*, 137–152.

Cucculelli, M., & Goffi, G. (2016). Does sustainability enhance tourism destination competitiveness? Evidence from Italian destinations of excellence. *Journal of Cleaner Production, 111*, 370–382.

Da Cunha Bezerra, M. C., Gohr, C. F., & Morioka, S. M. (2020). Organizational capabilities towards corporate sustainability benefits: A systematic literature review and an integrative framework proposal. *Journal of Cleaner Production, 247*, 119114.

Dangelico, R. M., Pontrandolfo, P., & Pujari, D. (2013). Developing sustainable new products in the textile and upholstered furniture industries: Role of external integrative capabilities: Developing sustainable new products. *Journal of Product Innovation Management, 30*(4), 642–658.

Denicolai, S., Cioccarelli, G., & Zucchella, A. (2010). Resource-based local development and networked core-competencies for tourism excellence. *Tourism Management, 31*, 260–266.

Di Bella, A., Petino, G., & Scrofani, L. (2019). The Etna macro-region between peripheralization and innovation: Towards a smart territorial system based on tourism. *Regional Science Policy & Practice, 11*, 493–507.

Dwyer, L., & Kim, C. (2003). Destination competitiveness: Determinants and indicators. *Current Issues in Tourism, 6*(5), 369–414.

Dyer, J. H., & Singh, H. (1998). The relational view: Cooperative strategy and sources of interorganisational competitive advantage. *The Academy of Management Review, 23*(4), 660–679.

EC, European Commission. (2020). *Compendium of best practices 2019 & 2020 European Capital of Smart Tourism competitions*. Scholz & Friends Agenda.

Eisenhardt, K. M., & Martin, J. A. (2000). Dynamic capabilities: What are they? *Strategic Management Journal, 21*, 1105–1121.

Evans, N. G. (2016). Sustainable competitive advantage in tourism organizations: A strategic model applying service dominant logic and tourism's defining characteristics. *Tourism Management Perspectives, 18*, 14–25.

Font, X., & McCabe, S. (2017). Sustainability and marketing in tourism: Its contexts, paradoxes, approaches, challenges and potential. *Journal of Sustainable Tourism, 25*(7), 869–883.

Foronda-Robles, C., Galindo-Pérez-de-Azpillaga, L., & Fernández-Tabales, A. (2020). Progress and stakes in sustainable tourism: Indicators for smart coastal destinations. *Journal of Sustainable Tourism*, 1–20.

Gill, A. M., & Williams, P. W. (2011). Rethinking resort growth: Understanding evolving governance strategies in whistler, British Columbia. *Journal of Sustainable Tourism, 19*(4–5), 629–648.

Godin, B., & Gaglio, G. (2019). How does innovation sustain 'sustainable innovation'? In F. Boons & A. McMeekin (Eds.), *Handbook of sustainable innovation* (pp. 27–37). Edward Elgar.

Gonçalves de Almeida, J.-M., Gohr, C.-F., Morioka, S.-N., & Medeiros da Nobrega, B. (2021). Towards an integrative framework of collaborative capabilities for sustainability: A systematic review and research agenda. *Journal of Cleaner Production, 279*, 123789.

González-Reverté, F. (2019). Building sustainable smart destinations: An approach based on the development of Spanish smart tourism plans. *Sustainability, 11*, 6874.

Graci, S. (2013). Collaboration and partnership development for sustainable tourism. *Tourism Geographies, 15*(1), 25–42.

Gutiérrez-Rodríguez, A. J., Barón, N. J., & Guaita Martínez, J. M. (2020). Validity of dynamic capabilities in the operation based on new sustainability narratives on nature tourism smes and clusters. *Sustainability, 12*, 1004.

Hassan, S. S. (2000). Determinants of market competitiveness in an environmentally sustainable tourism industry. *Journal of Travel Research, 38*, 239–245.

Haugland, S. A., Ness, H., Grønseth, B. O., & Aarstad, J. (2011). Development of tourism destinations: An integrated multilevel perspective. *Annals of Tourism Research, 38*(1), 268–290.

Hjalager, A. M. (2010). A review of innovation research in tourism. *Tourism Management, 31*, 1–12.

Inigo, E. A., & Albareda, L. (2019). Sustainability oriented innovation dynamics: Levels of dynamic capabilities and their path-dependent and self-reinforcing logics. *Technological Forecasting and Social Change, 139*, 334–351.

Jamrozy, U. (2007). Marketing of tourism: A paradigm shift toward sustainability. *International Journal of Culture, Tourism and Hospitality Research, 1*(2), 117–130.

Kemp, R., & Pearson, P. (2007). *Final report MEI project about measuring eco-innovation* (pp. 1–120). European Commission.

Kemper, J. A., & Ballantine, P. W. (2019). What do we mean by sustainability marketing? *Journal of Marketing Management, 35*(3-4), 277–309.

Lemmetyinen, A., & Go, F. M. (2009). The key capabilities required for managing tourism business networks. *Tourism Management, 30*, 31–40.

Leonidou, L. C., Leonidou, C. N., Fotiadis, T. A., & Zeriti, A. (2013). Resources and capabilities as drivers of hotel environmental marketing strategy: Implications for competitive advantage and performance. *Tourism Management, 35*, 94–110.

Pace, L. A. (2016). How do tourism firms innovate for sustainable energy consumption? A capabilities perspective on the adoption of energy efficiency in tourism accommodation establishments. *Journal of Cleaner Production, 111*, 409–420.

Pace, L. A., & Miles, I. (2020). The influence of KIBS-client interactions on absorptive capacity-building for environmental innovation. *European Journal of Innovation Management, 23*(4), 553–580.

Paulraj, A. (2011). Understanding the relationships between internal resources and capabilities, sustainable supply management and organizational sustainability. *Journal of Supply Chain Management, 47*(1), 19–37.

Paunovic, I., Dressler, M., Mamula Nikolic, T., & Popovic-Panti, S. (2020). Developing a competitive and sustainable destination of the future: Clusters and predictors of successful national-level destination governance across destination life-cycle. *Sustainability, 12*, 1–15.

Pechlaner, H., Herntrei, M., & Kofink, L. (2009). Growth strategies in mature destinations: Linking spatial planning with product development. *Tourism Review, 57*(3), 285–307.

Perles-Ribes, J. F., & Ivars-Baidal, J. (2018). Smart sustainability: A new perspective in the sustainable tourism debate. *Journal of Regional Research, 42*, 151–170.

Pikkemaat, B., Peters, M., & Bichler, F. (2019). Innovation research in tourism: Research streams and actions for the future. *Journal of Hospitality & Tourism Management, 41*, 181–196.

Prats, L., Guia, J., & Molina, F. X. (2008). How tourism destinations evolve: The notion of tourism local innovation system. *Tourism and Hospitality Research, 8*(3), 178–191.

Reinhold, S., Laesser, C., & Beritelli, P. (2015). 2014 St.Gallen Consensus on destination management. *Journal of Destination Marketing & Management, 4*, 137–142.

Rennings, K. (2000). Redefining innovation—Eco-innovation research and the contribution from ecological economics. *Ecological Economics, 2*(2), 319–332.

Rodríguez-Díaz, M., & Espino-Rodríguez, T.-F. (2008). A model of strategic evaluation of a tourism destination based on internal and relational capabilities. *Journal of Travel Research, 46*, 368–380.

Rodríguez-Díaz, M., & Espino-Rodríguez, T.-F. (2016). Determining the sustainability factors and performance of a tourism destination from the stakeholders' perspective. *Sustainability, 8*, 951.
Romão, J. (2020). Variety, smart specialization and tourism competitiveness. *Sustainability, 12*, 5765.
Saarinen, J. (2006). Traditions of sustainability in tourism studies. *Annals of Tourism Research, 33*(4), 1121–1140.
Sainaghi, R., De Carlo, M., & D'Angella, F. (2019). Development of a tourism destination: Exploring the role of destination capabilities. *Journal of Hospitality & Tourism Research, 43*(4), 517–543.
Seguí-Amortegui, L., Clemente-Almendros, J. A., Medina, R., & Grueso Gala, M. (2019). Sustainability and competitiveness in the tourism industry and tourist destinations: A bibliometric study. *Sustainability, 11*, 6351.
Sharpley, R. (2020). Tourism, sustainable development and the theoretical divide: 20 years on. *Journal of Sustainable Tourism, 28*(11), 1932–1946.
UNWTO, United Nations World Tourism Organisation. (2019). *Tourism definitions.*
Weidenfeld, A. (2018). Tourism diversification and its implications for smart specialisation. *Sustainability, 10*, 319.
Zehrer, A., Raich, F., Siller, H., & Tschiderer, F. (2013). Leadership networks in destinations. *Tourism Review, 69*(1), 59–73.
Zollo, M., Cennamo, C., & Neumann, K. (2013). Beyond what and why: Understanding organizational evolution towards sustainable enterprise models. *Organization & Environment, 26*(3), 241–259.

3

Sustainable Livelihoods and Tourism: An Overview of the Nexus in Developing Countries

Annarita Paiano, Miraj Ahmed Bhuiyan, Giovanni Lagioia, and Tiziana Crovella

Introduction

Livelihood is a key concept in grappling with poverty reduction. The term livelihood refers to the set of skills, assets (both material and social resources), and activities required for living. Livelihoods are sustainable if they can rebound from stresses and shocks and maintain or enhance capabilities and assets (current standard of living) without undermining the natural resource base on which it depends (DFID, 1999). The core values of livelihoods are people-centered, holistic (implemented in partnership), dynamic, build on the strengths of the impoverished, and link the "micro" with the "macro" in a sustainability focus. Nevertheless, in

A. Paiano (✉) • G. Lagioia • T. Crovella
Department of Economics, Management and Business Law, Largo Abbazia Santa Scolastica, University of Bari Aldo Moro, Bari, Italy
e-mail: annarita.paiano@uniba.it

M. A. Bhuiyan
School of Economics, Guangdong University of Finance and Economics, Guangzhou, China

T. Walker et al. (eds.), *Sustainable Tourism*,
https://doi.org/10.1007/978-3-031-43528-7_3

the past, studies on sustainable livelihoods have combined multiple meanings associated with this term. It has been sometimes associated with the concept of "resilience," and elsewhere compared to "vulnerability" towards forming a better understanding of livelihood determinant outcomes in a context of calamity, for instance (Phu & Chang-Richards, 2022).

Accordingly, Sustainable Livelihoods Approaches (SLA) emerged as a more effective and relevant concept towards poverty reduction through an understanding of poverty from the perspective of the poor. Initially conceived in the 1980s in the context of Farming Systems Research and Education, the approach was developed through the 1990s and crystallized as SLA in the late 1990s through the efforts of the Department for International Development (DFID) (Carney, 1998; Carney et al., 1999). Despite differences in the approach at a country level, several organizations have employed the Sustainable Livelihoods Approach and Framework in their development plans and programs. In particular, the framework has been used as a programming framework (UNDP) for programming analysis, design, monitoring, and evaluation (CARE Household Livelihood Security) and for integrating environmental sustainability (Carney et al., 1999). The Food and Agriculture Organization (FAO) of the United Nations has also built upon the SLA to find ways and means to improve the sustainable livelihoods of rural dwellers.

In this context, tourism emerges as a crucial economic sector: tourism business is considered one of the most common livelihood strategies for community development in developing countries (Shen et al., 2008). However, tourism development can involve positive and negative changes in host destinations' local livelihoods, which sometimes makes it difficult to achieve the Sustainable Development Goals (SDGs) due to impacts on the economy, society, and the environment.

This chapter aims to identify the operational and institutional variables that underpin the relationship between sustainable tourism and sustainable livelihoods in a given region. To this end, we propose an interdisciplinary framework, the Sustainable Livelihood Framework in Figure 3.1 (SLF, Fig. 3.1), which is based on a first elaboration by Chambers and Conway (1992) and adapted according to additional elements such as interdependence and functionality. This approach puts people at the center of development, thereby increasing the effectiveness

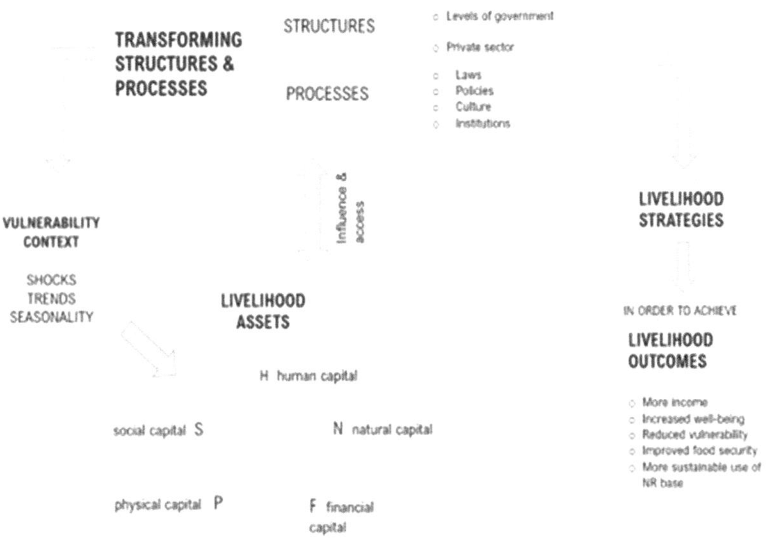

Fig. 3.1 Sustainable livelihood framework (SLF)

of development assistance. In this simple form, the framework involves people operating in a vulnerable context.

Beyond what the framework already considers, it has emerged that social, institutional, and organizational environments also influence livelihood strategies by combining and using assets. Using this observation as a starting point and considering Chambers and Conway's framework as well as our reworking, this new framework aims to:

- demonstrate how the livelihood approach fits the general objectives of the sustainable development goals,
- explain the livelihood framework,
- explain the links with existing / past approaches and methodologies,
- present suggestions on how to apply the approach concretely to tourism sustainability,
- identify priorities for future research and studies,
- identify significant items by reviewing projects, experience, and literature.

Subsequently, we contextualize this general framework of livelihood (SLF) in the the field of tourism (Fig. 3.2) using the scheme developed recently by Ming Su et al. (2019b). Figure 3.2 provides our own interpretation of the framework in the context of tourism based on Ming Su et al.'s earlier elaborations. Using the SDGs as a point of departure, we observe the tourism industry in terms of sustainable development, analyzing particular livelihoods in the context of a tourism industry closely interconnected with a given region.

In practice, the methodological path is based on several interconnected levels of analysis. First, we look at a country's dominant environmental perspective, the quality of life people enjoy, and the products and services the country provides. These three elements positively influence a tourist's choice of destination and where they spend both time and money. For this reason, we use indicators such as number of protected areas, level of implementation of Tourism Satellite Account, and System of Environmental and Economic tables in order to assess positive impacts.

Moreover, we use the Tourism Sustainable Development Index (TSDI), the Environmental and Climate Index (ECI), and the Health

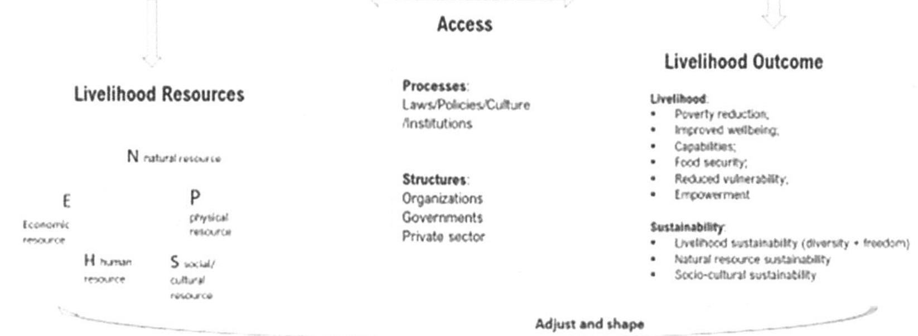

Fig. 3.2 Sustainable livelihood framework for tourism (SLFT)

Tourism Economic index (HTE) to investigate the sustainability of the system. The TSDI is a unique and complete indicator of sustainable development for the tourism sector. It overcomes data gaps and simultaneously allows data collection from heterogeneous and new data sources over time. The TSDI reflects the ratio of correlation between the economic development of tourism and its environmental impact. This indicator considers information on CO_2 emissions per capita per year, air quality, level of urbanization, vegetal forest cover, proximity to protected areas, education index, and life expectancy at birth (TSDI, 2022). The ECI provides information on a country's environmental health status, in particular the value of the CO_2 describes the level of the impacts of global greenhouse gas emissions and energy greenhouse gas emissions. Therefore, the ECI indicates the most critical pollution proxy to date. Finally, HTE represents the social and economic component, which aims to formulate normative judgments on countries to develop development programs (TSDI, 2022). Regarding statistical representation, with some exceptions, all indicator values (TSDI, ECI, and HTE) are standardized and can have a value between 0 and 10 (TSDI, 2022).

Generally, a country characterized by political polarization and a low measure of tolerance can negatively influence the overall perception of that location and downgrade the strength of the country's brand as a tourist destination. Furthermore, perceptions of a country's environmental compatibility can influence destination choice. Japan, for example, is one of the top 10 producers of CO_2 emissions worldwide but has also shown the most significant improvement in the perception of environmental compatibility (Future Brand, 2019). Therefore, this chapter adapts the SLF to the tourism context and applies it to critically evaluate the relationships between tourism and livelihoods to examine how these interactions affect sustainability. We then provide a replicable framework that extends to tourism and discuss the practical sustainbility implications of the livelihood of a given area.

The structure of this chapter is as follows: The first section "Literature Review" provides a review of existing literature and is composed of two subsections: sustainable livelihood in a tourism context and consequences of the COVID-19 pandemic. The following section, "Materials and Methods" introduces our methodology and clarifies variables considered, data, and indicators observed. In the "Results and Discussion" section, we present the results of the analysis and offer a discussion of the interactions between the different approaches we carried out. Finally, the last section, "Conclusions, Limitations, and Practical Implications" concludes this study by acknowledging its limits and pointing to future research directions.

Literature Review

Although tourism is promoted in rural regions of developing countries as a means of alleviating poverty and being a vector for modernization, few detailed case studies have analyzed its integration into media programs and considered its impact on impoverished segments of the population, particularly among national minorities (Cornet, 2015). Many developing countries imposed a temporary local emergency when the first case of infection from COVID-19 was recorded; this impacted the tourism industry severely and led to major shutdowns. In developed countries, such as those in Central Europe, there was a decrease in employment at tourist accommodation facilities in the period from March 2020 to June 2020 equal to about 85% compared to the same period in the previous year (Bartosz, 2021). In developing countries (e.g. parts of Northeast Asia and Southeast Asia, the Middle East, sub-Saharan Africa, and South Asia) there was a decrease of 77% of the total tourism traffic for the same period (Bartosz, 2022). This reduction also had repercussions on trade as imports and exports were slower or were interrupted. During this economic crisis that was unfolding parallel to the related health crisis, developing countries had to depend on humanitarian aid, which is not always adequate to mitigate the hardships of vulnerable populations. The transport sector stopped operations due to border closures or travel bans imposed in several countries, such as member countries of the European Union, the United States, and China, thus disrupting the supply chain of primary goods such as food (Reardon et al., 2020) and humanitarian aid donated by different agencies, such as the Food and Agriculture Organization (FAO) and the United Nations (UN).

Sustainable Livelihood in a Tourism Context

Tourism has long been considered an essential bridge to minimize the wealth gap between different population segments (Zhao and Ritchie, 2007). However, the industry has other side effects, such as the depletion of natural resources as Little and Blau (2020) highlight. Particularly, considering tourism as a strategy for overcoming the economic gap within and between countries, the sector must transition towards being fully sustainable from an economic, social, and environmental point of view. Therefore, if developing countries can grow thanks to tourism, simultaneously tourism must increase local revenues, create diversified jobs, guarantee greater food security, and reduce the depletion of natural resources. This can be achieved, for example, by avoiding the depletion of soil, deforestation, and water waste in regions such as Amazonia, Africa, and Asia. More specifically, the concept of tourism for the impoverished (PPT) has gained popularity (Akyeampong, 2011), and the approach to sustainable livelihoods (SL) has recently been adopted as an analytical tool in its exploration (Ashley, 2000; Petersen & Pedersen, 2010).

At its inception, the concept of sustainable livelihoods was mainly investigated in studies on rural communities before being extended to various analyses both in developed and developing countries (Wu & Pearce, 2012). As gestured to above, Chambers and Conway's (1992) definition of this framework comprises the skills, goods, and activities necessary to earn a living. So, a livelihood can be defined as being studied and conducted at the family level (basic level) and then combined, explicitly or implicitly, to provide insights at the community level (extended level) (Su et al., 2016c). Chambers and Conway (1992) also define livelihood as the ability to recover from stress or shock, while maintaining or enhancing the ability of resources to provide sustenance opportunities for subsequent generations. Furthermore, the authors point out that livelihoods are sustainable if communities can cope with change and generate positive effects on opportunities for future generations at different geographical and temporal scales.

Destination communities and their livelihoods are critical to tourism sustainability and regional development, as Rongna and Sun (2022) highlight. The livelihood challenges encountered by host populations include contexts of vulnerability, processes of transformation, as well as the temporal and spatial constraints in daily life (Rongna & Sun, 2020). However, to improve the sustainability of a livelihood, it is necessary to develop synergistic relationships between potential activities that impact on the territory on which the livelihood depends, such as tourism (Su et al., 2016c).

Usually, the Sustainable Livelihood Framework (SLF), as defined by the United Kingdom Department for International Development (DFID), is composed of five key components (Fig. 3.1): human capital, natural capital, financial capital, social capital, and physics capital. With the reconstruction of the SLF in the tourism context, according to Su et al. (2016a, 2016b, 2016c), various efforts were made better to represent tourism's role in the livelihood system. Various macro-conditions and trends define the broad context in which people and livelihoods interact with each other, underlined by Su et al. (2016a, 2016b, 2016c). For the inputs used to generate results in a livelihood assessment system, natural, physical, economic, human, social, and cultural resources were considered. Nonetheless, the cultural component plays a vital role in translating livelihood resources into livelihood strategies. This is a dynamic and interactive system, considering that livelihood outcomes can influence sustenance and can in turn influence the context and macro conditions that shape access to the resources that underpin sustenance (Su et al., 2016a, 2016b, 2016c). Hence, in the case of successful livelihood systems, a positive impacts are circular.

For this reason, the sustainable livelihood approach should be used in tourism development (Tao & Wall, 2009). Given the importance of sustainability in tourism development, the SLF has been used in previous studies as a theoretical framework for evaluating sustainable livelihood outcomes in tourism development. Tao and Wall (2009), for example, applied the SLF to the tourism context for the first time, analyzing the relationships between tourism and other livelihood strategies in an indigenous community in Taiwan. As for the developing countries, they must necessarily safeguard their vulnerabilities. Wageless, landless workers, and disenfranchised groups, including women and children as well as minority ethnic groups, must be protected in order to take action against events as impactful as global pandemic and to achieve the SDGs (Workie et al., 2020). A study based on the Sustainable Livelihoods Framework for Tourism (SLFT) analyzed the case of Aceh, Indonesia, a destination affected in 2004 by a severe natural disaster shows that the country subsequently developed tourism as a tsunami recovery strategy (Bingjie et al., 2020). The responses of the tourism industry in the post-disaster phase are highly variable, and some destinations have difficulty attracting visitors even if they have completed reconstruction and restoration (UN, 2005). On the other hand, other places have become more attractive for tourists who visit sites linked to disasters and with a strong emotional impact (Miller, 2008). This type of tourist travel is a new phenomenon known as "post-disaster dark tourism" (Prayag, 2016).

Consequences of the COVID-19 Pandemic

Developing countries disproportionately experienced the consequences of the COVID-19 pandemic on health, livelihoods, and tourism (Workie et al., 2020). The COVID-19 pandemic also brought fear to "consumer countries" regarding global food crises, such as in 2008 (World Bank, 2020). Due to the impacts of the pandemic on trade, exports, and imports, alongside the invasion of Ukraine (beginning February 2022), global food security is at risk with a particular effect on developing countries. Export restrictions and sanctions resulting from the Russia-Ukraine conflict have put global food security at risk, particularly in developing countries that import food exclusively from developed countries. As underlined by Soliku et al. (2021), the major socio-economic impact of the COVID-19 pandemic has registered in terms of loss of livelihoods. Furthermore, in rural areas, the means of subsistence have a direct connection with tourism, as it is the primary source of livelihood in developing and non-industrialized areas. For this reason, livelihoods in the form of tourism are believed to generate more income and at a faster rate, especially in areas with delayed development.

Other authors, such as Sisneros-Kidd et al. (2019), highlight that tourism, on the one hand, serves as a source of income for many rural communities; on the other hand, the excessive dependence on tourism can negatively influence communities by preventing their ability to diversify. Therefore, the effects of the pandemic on tourism are manifold: falling incomes, business closures, or losses of jobs and other forms of livelihood, as highlighted by Elo et al. (2014). Finally, according to UNWTO estimates, 2020 was the worst year for tourism, as international arrivals fell by 73%. Based on some provisional data, the number of international arrivals in December 2020 was 65% lower than in 2019. Notably, the full impact of emerging variants cannot be said to be over yet, especially in developing areas with moderate subsistence levels. This rise in cases is also related to tourist travel as well.

For this reason, the pace of recovery remains slow and uneven (based on different levels of mobility restrictions, vaccination rates, and traveler confidence). For example, the Caribbean recorded the best performance in 2021 (+63% above 2020). In 2022, only a few destinations were approaching or exceeding pre-pandemic levels. Africa recorded a 74% decline in 2020 compared to 2019. In the Middle East, arrivals fell by 79% in 2020 compared to 2019. In Asia and the Pacific, arrivals were still below 94% compared to pre-pandemic values, as many destinations remained closed to leisure travel (UNWTO, 2022).

Materials and Methods

With the aim of analyzing sustainable tourism development, we proceed by identifying, first, the various indicators associated with the topic. In particular, those indicators are identified in an interdisciplinary manner representing tourism's effects on sustainability and human subsistence in developing countries. As previously mentioned, the significance of this integrated observation approach (livelihoods, tourism, sustainable development) is also satisfied in the United Nations 2030 Agenda. In particular, in the Sustainable Development Goals. Goal 15, target 15.C, for example, supports the growing capacity of local communities to engage in sustainable livelihood opportunities. Furthermore, the link between the approach to livelihood and sustainable tourism also immediately emerges in goal 12, target 12.B.

With this in mind, we propose a list of sustainability indicators based on the above literature review. Then, given that the Sustainable Livelihoods Framework (SLF) has often been used as a strategic guideline to guide community development (Brocklesby & Fisher, 2003; Serrat, 2017), we identify four critical components of this process (Arce, 2003; Brocklesby & Fisher, 2003). The first component is the vulnerable context: this suggests that the stability of communities can be affected by various problems such as shocks, seasonality, and critical trends. Therefore, these events can damage the heritage of communities (Su et al., 2016a, 2016b). The second component is the capital patrimony that a community possesses. There are five types: local human, social, natural, financial/economic, and physical capital. The third component is the livelihood strategies, commonly implemented to develop a community and contribute to its capital. Finally, the fourth component is represented by the policies and institutional support necessary to maximize the effectiveness of strategic livelihood choices. We propose advancing the basic SLF framework into SLFT, considering these significant variables.

Variables and Methodology

The analysis of methodologies gathered from the scientific literature is not univocal, because different methodologies have been used, as Cernat and Gourdon (2012) underline. Asmelash and Kumar (2019) claim that very little progress has been made in monitoring and measuring sustainable tourism development, using integrated approaches with interpretations such as SDGs and livelihoods.

A study by Shammi et al. (2020) highlights demographic characteristics such as the health conditions of the population, the health system, governance, and political issues, government decisions and impacts, socio-economic issues, immediate emerging issues (such as tourism and pandemic), and permanent environmental and social problems. Particularly, these scholars consider these elements as variables affected by, and that can influence, tourism. Consequently, in order to address these issues, we suggest the use of TSDI, ECI, and the indexes outlined in the introduction.

Furthermore, based on the proposed analysis, we additionally analyze a country's perception of the relationship between the COVID-19 pandemic effects and the livelihoods in a generic tourist area characterized by different levels of environmental performance. Figure 3.3 displays the proposed methodological framework, aiming to fill the gap in the existing literature.

We begin with an analysis of livelihoods through the evaluation of a country's budget plans to achieve economic, social, and environmental

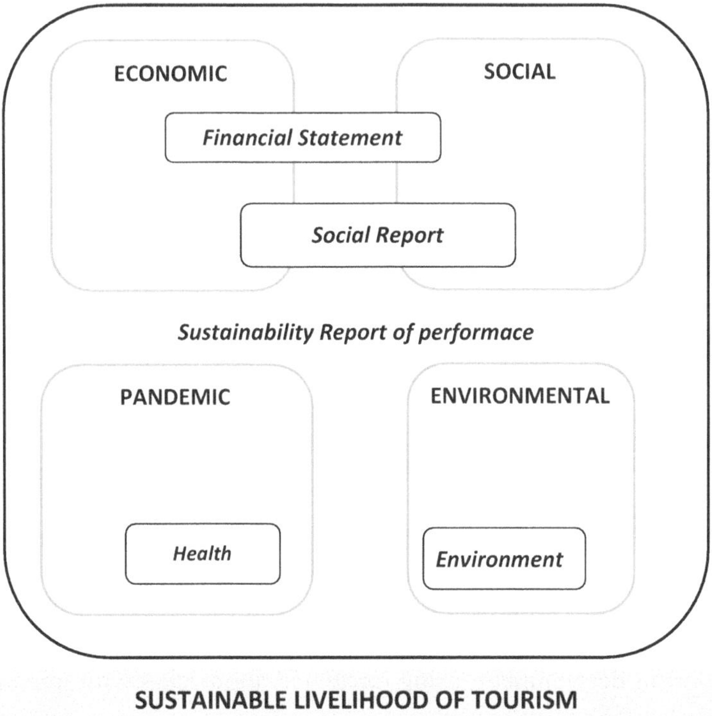

Fig. 3.3 Sustainable livelihood framework for tourism post COVID-19

sustainability in order to understand the pandemic's effects on sustainability levels and its impact on the target country. A combination of qualitative and quantitative methods can be used to build indicators and study the impact of COVID-19 on tourism in a sample area. For this reason, to analyze the relationship between tourism and sustainable livelihoods in the context of the impact of COVID-19 in a sample area, we provide criteria, illustrated in Table 3.1. Also in Table 3.1, we provide a matrix to guide observations of the relationship between tourism, livelihoods, and sustainable development after a catastrophic event such as a pandemic.

The observations that Table 3.1 presents can be based on an overview of the values associated with the Global Reporting Initiatives (GRI). These are reporting standards developed by GSSB (2020), and widely

Table 3.1 Macro-variables for sustainable livelihood and tourism

Macro-variables	Indicators	Key elements
Universal information	Population	
	Territory	
	Transport	
	Level of education	
	Health and policies	
Specific information	Economic	Economic performance (tourism income, agricultural income, industrial income)
		Market presence
		Indirect economic impact
		Procurement impact
		Anti-corruption
		Tax
	Environmental	Materials
		Energy
		Water and effluents
		Biodiversity
		Emissions
		Waste
		Environmental compliance
		Suppliers environmental assessment
	Social	Employment (local employment and migration work)
		Labor/management relations
		Occupational health and safety
		Training and education
		Diversity and equal opportunity
		Non-discriminations
		Child labor
		Forced and compulsory labor
		Security practices
		Human right assessment
		Local communities
COVID-19	Number of deaths	
	Sectors in lockdown	
	Gross domestic product	

Source: Authors' elaboration

used by reporting companies and countries, which also observe the Sustainable Livelihood Framework in a tourism context among the proposed variables (Fig. 3.2). The variables contained in the GRI correspond to the drivers contributing to the mapping of sustainable tourism as an extension of responsible tourism, ecotourism, CSR, local ownership, and policymakers. In this context, these drivers constitute a roadmap for the tourism sector toward sustainable development, defining the priority material areas of intervention in a long-term perspective. The goal is to ensure responsible innovation and the livelihood of the tourism sector and the wider tourist community, especially in a post-COVID-19 situation. We present the macro indicators to be considered in Table 3.2.

In addition, we present a list of variables concerning tourism, businesses, and other key stakeholders in the destination management perspectives in Table 3.3.

This systematic interpretation and application of variables and indicators has been adapted with the aim of providing the snapshot of the relationship between tourism, sustainable development, and livelihoods.

Table 3.2 The hierarchy of tourism sustainability assessment

System of touristic destination	Human system	Economic sustainability	Employment quality
			Economic viability
			Local prosperity
		Socio-cultural sustainability	Social equity
			Visitor fulfilment
			Local control
			Community well-being
			Cultural richness
		Institutional sustainability	Local-oriented control policy
			Political participation
			Local policy planning
			Political support
	Ecosystem	Environmental sustainability	Physical integrity
			Biological diversity
			Resources efficiency
			Environmental purity

Source: Authors' elaboration adapted from Asmelash and Kumar (2019)

Table 3.3 Evaluation parameters of a destination's management

Baseline aspects	Indicators
Energy management	Energy consumption attributable to tourism
	Percentage of renewable energy consumption with respect to the total for tourism
Water management	Water consumption attributed to tourism
Wastewater management	Volume of reused water
	Volume of treated wastewater
	Volume of wastewater
Atmospheric pollution	Daytime noise levels
	Night-time noise levels
	Pollutant emissions levels
Intensity of tourist use	Total tourists per unit area
Economic benefits of tourism for the host community and destination	Total number of tourist arrivals
	Average length of stay
	Tourism revenues
Sustaining tourist satisfaction	Global satisfaction level of tourist
	Evaluation of the price-quality relationship by tourist
Seasonality of tourism activity	Percentage of official tourism accommodation establishments that are open all year
Tourism employment	Total number of individuals employed in the tourism sector
Destination competitiveness	Average occupancy rate for official tourism accommodation establishments
Socio-cultural effects of tourism on host community	Health care equipment
	Number of passenger transport vehicles per inhabitant
	Number of services sector establishments per inhabitant
	Number of pharmacies per inhabitant
Local public safety	Evaluation of destination safety by tourists
	Number of accidents involving fatalities on urban roads per 1,000 persons
Conservation of cultural heritage	Number of protected designated sites
	Pressures of cultural heritage
Effects of local population structure	Variation of population level
	Percentage of young population
	Number of individuals per unit destinations area

(*continued*)

Table 3.3 (continued)

Baseline aspects	Indicators
Social carrying capacity of the destination and effect on well-being in the local population	Percentage of foreign population Percentage of population enrolled in non-compulsory education General demographic dependency index

Source: Authors' elaboration adapted from Lozano-Oyola et al. (2019)

Table 3.4 Number of protected areas in the world

	Areas	Number of protected areas
1	Asia and Pacific	35,652
2	Africa	9114
3	Europe	162,401
4	Latin America and the Caribbean	10,039
5	Polar	34
6	North America	52,958
7	West Asia	380

Source: Authors' elaboration data Protected Planet (2022)

Data and Indicators Observed

We collect data, samplings, and observations regarding protected areas, the level of implementation of the Tourism Satellite Account, the Environmental System, and the Income Statement based on the needs outlined in the above section. Moreover, we evaluate the TSDI, ECI, and HTE indices. Firstly, the number of protected areas (Table 3.4) represents the leading indicator of the "World Database on Protected Areas (WDPA)": it is a global and more complete quantification database of marine protected areas present on the planet (Protected Planet, 2022). This is an indicator confirmed by the United Nations in the program for the environment, the International Union, and the conservation of nature (IUCN). Secondly, Table 3.5 shows the level of implementation of "standards accounting tools" to monitor tourism's effects on the livelihoods from economic and environmental points of view. Specifically, we have considered the 11 tables of accounts (Tourism Satellite Account-TSA and System of Environmental and Economic Accounting-SEEA)

Table 3.5 Level of implementation of Tourism Satellite Accounts and System of Environmental and Economic Accounts

	Areas	Level of implementation
1	Asia and Pacific	Lowest/medium
2	Africa	Lowest
3	Europe	Lowest/medium
5	Oceania	Highest
6	North America	Medium/highest
7	South America	Lowest/medium

Source: Authors' elaboration on data Our World (2022)

deemed most relevant and feasible for monitoring sustainability in tourism. The TSA we consider are:

1. TSA Table 1 on inbound tourism expenditure,
2. TSA Table 2 on domestic tourism expenditure,
3. TSA Table 3 on outbound tourism expenditure,
4. TSA Table 4 on internal tourism expenditure,
5. TSA Table 5 on production accounts of tourism industries,
6. TSA Table 6 domestic supply and internal tourism consumption,
7. TSA Table 7 on employment in tourism industries,
8. SEEA Table water flows,
9. SEEA Table energy flows,
10. SEEA Table GHG emissions,
11. SEEA Table solid waste (UNSTATS, 2022).

Finally, through the TSDI index (Table 3.6), known as the adjustment of the development index applied to the tourism sector, we observe the relationship between tourism, environment, climate, and health. This new generation metric includes the pressure of biodiversity, the pressures that govern the social and economic components and allows us to summarize the relationship between means of subsistence, tourism, and development (TSDI, 2022). The matching of these indicators allows for the formulation of nuanced observations on countries and proposes support measures to assist them in improving their environmental strategies. In particular, the TSDI index engages destinations to become more sustainable places to visit. From a methodological point of view, the TSDI,

Table 3.6 Relationship between the TSDI, ECI, and HTE indices

	Countries worldwide ranking	TSDI	ECI	HTE
1	Costa Rica	10.0	10.0	3.82
2	Sweden	7.17	9.46	4.68
3	Slovenia	6.63	9.46	4.33
4	Slovakia	6.54	9.59	3.83
5	New Zealand	6.46	9.46	4.22
6	Japan	6.2	8.14	8.19
7	Malaysia	5.87	8.63	6.32
8	Bahamas	5.4	9.46	3.52
9	France	5.24	7.0	10.0
10	Fiji	5.08	9.59	2.96
11	Spain	4.92	7.14	9.04
12	UK	4.92	7.44	8.28
13	Croatia	4.74	8.28	5.94
14	Austria	4.40	7.44	7.40
15	Dominica	4.32	9.90	1.84
16	Hungary	4.27	7.88	6.22
17	Finland	4.11	8.76	4.14
18	Germany	4.11	6.52	8.84
19	Italy	3.93	6.17	9.14
20	Denmark	3.71	7.57	5.98

Source: Authors' elaboration on data TSDI (2022)

based on reliable data, allows for the improvement of strategy and guides a sustainable decision-making process for tourism (TSDI, 2022).

Finally, the following section contains the results and a critical discussion on the monitoring carried out by applying the framework of Sustainable Livelihood on tourism and several indicators confirmed by the United Nations in the program for the environment, the International Union, and the conservation of nature.

Results and Discussion

Starting from the definition of livelihoods, defined as "the skills, material and social resources and activities required for a means of subsistence" (Scoones, 1998, p. 5), we consider sustainable development is when "countries can cope with and recover from stress and recorded shocks (such as a pandemic or natural catastrophe) and maintain or improve their capabilities and resources both now and in the future without compromising the natural resource base" (Serrat, 2017, p. 21). Developing from these definitions and arguments, Shen et al. (2008) suggest that the direct application of the SLF to tourism is not enough because tourism must be treated in a context, integrating the SLF with tourism development, taking into account all the peculiar characteristics of tourism, confirming our thesis. In line with the SLF, the SLFT framework presents the interaction between different kinds of resources: human, social, natural, economic, and institutional capital (Fig. 3.4) in hierarchical relationships between variables (Table 3.2). Thus, if capital allows people to access tourism markets, share the benefits of tourism, access and participate in decision-making, and are reflected in political decisions (Shen et al., 2008), we identify a sustainable tourism livelihood. Thus, by observing the interaction among several macro-variables (Table 3.1), quantifying the different indicators (Table 3.3), and summarizing the results in several key points (Tables 3.4, 3.5, and 3.6), we have modeled a replicable methodological framework to clarify this interdependence between tourism, sustainable development, livelihoods, and pandemic influence as shown in Fig. 3.4.

For example, for a developing country such as Indonesia, which underwent years of war, an earthquake, and a tsunami, there is a clear decrease in visitors compared to the year before the tsunami.

To help developing countries recover from a tsunami, local and central governments have collaborated with numerous humanitarian organizations and foreign countries, devoting efforts to post-tsunami reconstruction through various projects (Jauhola, 2013). These efforts highlight tourism development as a critical initiative (World Tourism Organization UNWTO, 2005). Therefore, in terms of SDG indicator 15.C, we analyze protected areas overlay by Biodiversity Index, an indicator consisting of three sub-indicators: (a) the degree of protection of the world's terrestrial and marine ecoregions; (b) the degree of protection of the Important Bird and Biodiversity Areas (IBAs); and (c) the degree of protection of the Alliance for Zero Extinction (AZE) sites (SDSN, 2022). Mainly used on a global scale, this indicator can be calculated for regions, countries, or even biomes. In quantitative terms, Table 3.4 highlights the number of protected areas based on World Database on Protected Areas (WDPA) to present the results associated with SDG 15.C. Europe, with the largest

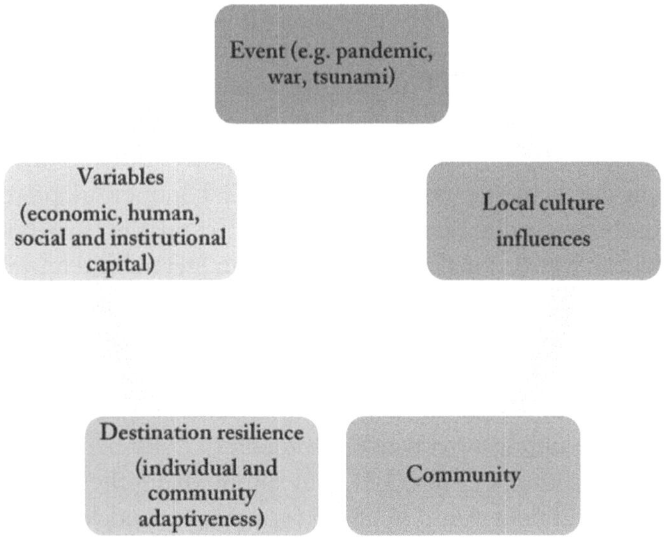

Fig. 3.4 Interaction among variables and SLFT

number of protected areas, offers tourists a chance to experience a sector that is attentive to the conservation of the natural environment, also thanks to the many European directives in force.

Conversely, to evaluate the relationship with SDG 12.B we monitor the economic and environmental aspects of tourism. Therefore, Table 3.5 shows the degree of implementation of the 11 Tourism Satellite Account and System of Environmental and Economic tables (listed in the Materials and Methods section), deemed most relevant and feasible for monitoring sustainability in tourism.

Oceania is currently the most sustainable continent, with a high implementation of Tourism Satellite Account, Environmental System, and Income statement tables. To research the relationship between tourism, sustainable development (especially in terms of achieving targets 12.B and 15.C of the SDGs), and livelihoods, we take into consideration the TSDI, ECI, and HTE (Table 3.6). These TSDI, ECI, and HTE are scored between 0 and 10%, with Costa Rica recording the best performance. Mainly, the TSDI aims to improve sustainability development strategy at the destination level. Its purpose is to enable destinations to become more resilient and sustainable places to visit, and currently, it is a sustainable development indicator unique to the tourism industry. It is a resilient indicator that does not present data gaps and is flexible because it welcomes heterogeneous and new data sources over time. Furthermore, it represents the most significant indicator from a scientific point of view and reflects the correlation between the economic development of tourism and its environmental impact.

Conclusions, Limitations, and Practical Implications

The Sustainable Development Goals (SDGs) were adopted by the United Nations in September 2015 through the plan "Agenda 2030" with the aim to represent an ambitious path for a global development towards the worldwide sustainability.

However, integrating the SDGs into the original framework of sustainable livelihood presents many limitations. Particularly, the use of these 17 SDGs seems to have run aground in reporting with respect to the mapping of the goals and indicators achieved, undermining a true integrative approach. This represents one of the major obstacles to the implementation of the SDGs in practical and managerial approaches. Moreover, approaches to livelihood need analytical reflection to examine a more politically informed approach, linking the analysis of location-specific livelihoods to broader perspectives of structural political economy. Specifically, stakeholders involved in sustainability issues should develop appropriate approaches considering external events, such as pandemics, wars, religious conflicts and any external element that can influence the path to achieving sustainability objectives. These considerations have to push towards the use of robust indicators and databases with the aim to provide clear and truthful observations.

Furthermore, through the observations carried out in this chapter and on the basis of the different indicators considered, the authors documented divergent performances. For example, if European countries presents the largest number of protected areas, this does not mean that they have the best level of implementation of economic and environmental accounts. Hence, considering the SDGs structure, all observations and quantified values demonstrated that a country can achieve full implementation of the targets of one goal but does not reach those of a complementary one. Then, to resolve this divergence, countries must commit to eliminating regulatory barriers and promoting greater interdependence between various economic sectors in order to encourage the close interdependence between the various goals.

References

Akyeampong, O. A. (2011). Pro-poor tourism: Residents' expectations, experiences and perceptions in the Kakum National Park Area of Ghana. *Journal of Sustainable Tourism, 19*(2), 197–213.

Arce, A. (2003). Value contestations in development interventions: Community development and sustainable livelihoods approaches. *Community Development Journal, 38*(3), 199–212.

Ashley, C. (2000). *The impacts of tourism on rural livelihoods: Namibia's experience*. Overseas Development Institute (ODI).

Asmelash, A., & Kumar, G. S. (2019). Assessing progress of tourism sustainability: Developing and validating sustainability indicators. *Tourism Management, 71*, 67–83. https://doi.org/10.1016/j.tourman.2018.09.020

Bartosz, K. (2021). The impact of political decisions on the tourist accommodation occupancy—Central Europe in the time of the COVID-19 pandemic. *Geographia Cassovensis, 15*(1), 27–36. https://doi.org/10.33542/GC2021-1-02

Bartosz, K. (2022). Impact of the COVID-19 pandemic on international tourism income in tourism receiving countries. *Sustainability, 14*(19), 12550. https://doi.org/10.3390/su141912550

Bingjie, L. L., Dini, M., Xiaoyuan, T., & Tianyu, Y. (2020). Can post-disaster tourism development improve destination livelihoods? A case study of Aceh, Indonesia. *Journal of Destination Marketing & Management, 18*, 100510. https://doi.org/10.1016/j.jdmm.2020.100510

Brocklesby, M. A., & Fisher, E. (2003). Community development in sustainable livelihoods approaches: An introduction. *Community Development Journal, 38*(3), 185–198.

Carney, D. (1998). *Sustainable rural livelihoods: What contribution can we make?* DFID.

Carney, D. et al. (1999) *Livelihoods approaches compared*. DFID. Retrieved May 15, 2020, from www.livelihoods.org/info/docs/lacv3.pdf

Cernat, L., & Gourdon, J. (2012). Paths to success: Benchmarking cross-country sustainable tourism. *Tourism Management, 33*, 1044–1056. https://doi.org/10.1016/j.tourman.2011.12.007

Chambers, R., & Conway, G. (1992). *Sustainable rural livelihoods: Practical concepts for the 21st century*. Institute of Development Studies (UK).

Cornet, C. (2015). Tourism development and resistance in China. *Annals of Tourism Research, 52*, 29–43.

DFID. (1999). *Strategic conflict and conflict impact assessment*. DFID.

Elo, S., Kääriäinen, M., Kanste, O., Polkki, T., Utriainen, K., & Kyngas, H. (2014). Qualitative content analysis: A focus on trustworthiness. *SAGE Open, 4*(1), 2158244014522633.

Future Brand. (2019). *Future brand country index, 2019. 150*. Retrieved May 28, 2022, from https://www.futurebrand.com/uploads/FCI/FutureBrand-Country-Index-2019.pdf

GSSB (Global Sustainability Standards Board). (2020). Consolidated set of GRI sustainability reporting standards 2020. Retrieved June 11, 2020, from https://www.globalreporting.org/standards/gri-standards-download-center/?g=1e908ae5-119a-4d3b-98e2-4320f73d5b3a

Jauhola, M. (2013). *Post-tsunami reconstruction in Indonesia: Negotiating normativity through gender mainstreaming initiatives in Aceh*. Routledge.

Little, M. E., & Blau, E. (2020). Social adaptation and climate mitigation through agritourism: A case study of tourism in Mastatal, Costa Rica. *Journal of Ecotourism, 19*(2). https://doi.org/10.1080/14724049.2019.1652305

Lozano-Oyola, M., Blancas, F. J., González, M., & Caballero, R. (2019). Sustainable tourism tags to reward destination management. *Journal of Environmental Management, 250*, 109458. https://doi.org/10.1016/j.tourman.2018.09.020

Miller, S. D. (2008). Disaster tourism and disaster landscape attractions after Hurricane Katrina: An auto-ethnographic journey. *International Journal of Culture, Tourism and Hospitality Research, 2*(2), 115–131.

Our World. (2022). Monitoring of the economic and environmental aspects of tourism, 2019. Retrieved May 25, 2022, from https://ourworldindata.org/grapher/implementation-of-tools-to-monitor-economic-and-environmental-tourism?time=latest&region=Oceania&country=AIA~KOR~BGD

Petersen, E. K., & Pedersen, M. L. (2010). *The sustainable livelihoods approach: From a psychological perspective: Approaches to develop*. Institute of Biology, University of Aarhus.

Phu, G., & Chang-Richards, A.-Y. (2022). Livelihood recovery for sustainable development: A study of the Lushan earthquake. *International Journal of Disaster Risk Reduction, 82*, 103317. https://doi.org/10.1016/j.ijdrr.2022.103317

Prayag, G. (2016). It's not all dark!: Christchurch residents' emotions and coping strategies with dark tourism sites. In C. M. Hall, S. Malinen, R. Vosslamber, & R. Wordsworth (Eds.), *Business and post-disaster management* (pp. 155–166). Routledge.

Protected Planet. (2022). World Database on Protected Areas (WDPA). Retrieved May 25, 2022, from https://www.protectedplanet.net/en/thematic-areas/wdpa?tab=WDPA

Reardon, T., Mishra, A., Nuthalapati, C., Bellemare, M., & Zilberman, D. (2020). Covid-19's disruption of India's transformed food supply chains. *Economic Political Weekly, 55*, 18–22.

Rongna, A., & Sun, J. (2020). Integration and sustainability of tourism and traditional livelihood: A rhythm analysis. *Journal of Sustainable Tourism, 28*(3), 455–474.

Rongna, A., & Sun, J. (2022). Tourism livelihood transition and rhythmic sustainability: The case of the Reindeer Evenki in China. *Annals of Tourism Research, 94*(103381), 1–14.

Scoones, I. (1998). Sustainable rural livelihoods: A framework for analysis (Working Paper, Institute of Development Studies). https://opendocs.ids.ac.uk/

Serrat, O. (2017). *The sustainable livelihoods approach*. Asian Development Bank.

Shammi, M., Bodrud-Doza, Reza Md, A., & Towfiqul, I. (2020). COVID-19 pandemic, socioeconomic crisis and human stress in resource-limited setting. A case from Bangladesh. *Heliyon, 6*(5), e04063, 1–12. https://doi.org/10.1016/j.heliyon.2020.e04063

Shen, F., Hughey, K. F., & Simmons, D. G. (2008). Connecting the sustainable livelihoods approach and tourism: A review of the literature. *Journal of Hospitality and Tourism Management, 15*(1), 19–31.

Sisneros-Kidd, A. M., Monz, C., Hausner, V., Schmidt, J., & Clark, D. (2019). Nature-based tourism, resource dependence, and resilience of Arctic communities: Framing complex issues in a changing environment. *Journal of Sustainable Tourism, 27*(8), 1259–1276.

Soliku, O., Kyiire, B., Mahama, A., & Kubio, C. (2021), Tourism amid COVID-19 pandemic: Impacts and implications for building resilience in the eco-tourism sector in Ghana's Savannah region. *Heliyon, 7*(9), e07892, 1–10.

Su, M. M., Wall, G., & Xu, K. J. (2016a). Heritage tourism and livelihood sustainability of a resettled rural community: Mount Sanqingshan World Heritage Site, China. *Journal of Sustainable Tourism, 24*(5), 735–757.

Su, M. M., Wall, G., & Xu, K. J. (2016b). Tourism-induced livelihood changes at mount Sanqingshan world heritage site. *China. Environmental Management, 57*(5), 1024–1040. https://doi.org/10.1007/s00267-016-0672-8

Su, M. M., Wall, G., & Jin, M. (2016c). Island livelihoods: Tourism and fishing at long islands, Shandong Province, China. *Ocean & Coastal Management, 122*, 20–29. https://doi.org/10.1016/j.ocecoaman.2015.11.014

Su, M. M., Sun, Y. H., Jiao, W. J., & Min, Q. W. (2019a). A community livelihood approach to agricultural heritage system conservation and tourism development: Xuanhua grape garden urban agricultural heritage site, Hebei Province of China. *Sustainability, 10*(2), 361.

Su, M. M., Wall, G., Wang, Y., & Jin, M. (2019b). Livelihood sustainability in a rural tourism destination—Hetu Town, Anhui Province, China. *Tourism Management, 71*, 272–281. https://doi.org/10.1016/j.tourman.2018.10.019

Sustainable Development Solutions Networks. (2022). *Protected areas overlay with biodiversity*. Indicators and a Monitoring Framework, Launching a data revolution for the Sustainable Development Goals. https://indicators.report/indicators/i-87/

Tao, T. C., & Wall, G. (2009). Tourism as a sustainable livelihood strategy. *Tourism Management, 30*(1), 90–98.

TSDI. (2022). TSDI Index. Retrieved May 23, 2022, from https://www.tourism-sdi.org/

UN agency seeks to boost tourism to tsunami-hit nations. (2005, March, 2). UN News Global perspective Human Stories. Retrieved from https://news.un.org/en/story/2005/03/130402-un-agency-seeks-boost-tourism-tsunami-hit-nations

UNSTATS. (2022). *SDG indicator metadata*. https://unstats.un.org/sdgs/metadata/files/Metadata-12-0b-01.pdf

UNWTO. (2022). *Impact assessment of the COVID-19 outbreak on international tourism*. https://www.unwto.org/impact-assessment-of-the-covid-19-outbreak-on-international-tourism

UNWTO. (2005). Making Tourism More Sustainable - A Guide for Policy Makers, UNEP and UN Tourism, 2005, p.11–12, https://www.unwto.org/sustainable-development

Workie, E., Mackolil, J., Nyika, J., Ramadas, S. (2020). Deciphering the impact of COVID-19 pandemic on food security, agriculture, and livelihoods: A review of the evidence from developing countries. *Current Research in Environmental Sustainability, 2*, 100014.

World Bank. (2020). Food Security and COVID-19. World Bank Brief. https://www.worldbank.org/en/topic/agriculture/brief/food-security-and-covid-19

Wu, M.-Y., & Pearce, P. L. (2012). Tourism research in and about Tibet: Employing a system for reviewing regional tourism studies. *Tourism and Hospitality Research, 12*(2), 59–72.

Zhao, W., & Ritchie, J. R. B. (2007). Tourism and poverty alleviation: An integrative research framework. *Current Issues in Tourism, 10*(2–3), 119–143.

4

Innovations in Luxury Tourism: Developing a Novel Theoretical Framework

Vasco Santos, Carlos Costa, Manuel Au-Yong, and Maria Sousa

Introduction

Sustainable tourism is defined by the United Nations World Tourism Organization (UNWTO) as being linked to the environmental, economic, and socio-cultural aspects underlying tourism development. In addition, as a way to mitigate the negative effects of tourism, the UNWTO announced that 2017 would be the year of sustainable tourism, taking into account the UN's sustainable development goals (SDGs) as applied to tourism, namely, the UNWTO Journey to 2030.

V. Santos (✉)
Higher Institute of Management and Administration of Santarém, Santarém, Portugal

C. Costa • M. Au-Yong
University of Aveiro, Aveiro, Portugal
e-mail: ccosta@ua.pt

M. Sousa
University Institute of Lisbon, Lisbon, Portugal

T. Walker et al. (eds.), *Sustainable Tourism*,
https://doi.org/10.1007/978-3-031-43528-7_4

The twenty-first century has seen a change in the meanings and motivations associated with luxury consumption with regard to the senses; the consumption of luxury products is much more focused on experiences, emotions, and pleasure (Machado & Sousa, 2018) than on the purchasing of objects. The ambivalent concept of luxury is often less of an individually defined concept and more often one constructed by societal pressures and impressions, though exceptions are notable in cases of personal achievement narratives (Correia et al., 2020). One could, for example, ponder the worth of climbing Mount Kilimanjaro (an example of adventure tourism) in terms of achievement and self-actualization. The help and support needed (e.g., guide, cook, bearer/carrier of equipment) in order for the average citizen (e.g., not a professional athlete or mountain climber) to climb the 5800 meters make it a luxury endeavor (due to the costs involved). The takeaway from this adventure is certainly more emotional and experiential than material (a few souvenirs brought home will not constitute the luxury experience). The whole experience may even involve months of preparation (e.g., physical preparation, special diet), which add to the satisfaction.

Poelina and Nordensvard (2018, p. 158) state that "sustainable luxury tourism might not just be desirable, but also necessary." However, to date, very little is known about how sustainability in the luxury tourism sector should operate in terms of the value chain, which concerns the functional and operational performance of the tourism and hospitality industry. The major gap is represented by the absence of scientific advancement in sustainable luxury tourism worldwide. This highlighted gap is reinforced by Moscardo (2017, p. 164), who attests that "given that there are a number of forces encouraging growth in luxury tourism and that there is an increasing number of companies claiming to provide sustainable luxury tourism opportunities, it seems timely to critically examine this concept." In fact, "sustainability issues have become a challenge for the luxury sector whose continuous success and growth—despite recessions, economic crises and increasing social inequality—appear as a paradox or even a provocation for some critics" (Kapferer & Michaut, 2015, p. 1). Within this context, consumers will, for example, react positively during room bookings when they know, through the marketing communication strategy of the hotel, that a luxury hotel is committed to sustainability, due to the fact that this information increases their positive perception of the integrity of the hotel (Amatulli et al., 2021).

Moscardo and Benckendorff further claim that "the emergence of concepts like 'intelligent' and 'sustainable' luxury raises a number of research questions, especially in terms of understanding consumer motivations and evaluations of luxury and sustainable experiences" (Moscardo & Benckendorff, 2010, p. 709). Overall, these authors highlight issues facing the longer-term sustainability of luxury tourism experiences. They present some evidence, though limited, of the emergence of tourist concerns about the environment and links between luxury and sustainable practices, especially for the Bed and Breakfast (B&B) category.

Sustainable luxury should aim to deliver high social and environmental performance, with a level of excellence across the entire value chain of organizations, as well as local cultural content (e.g., craftsmanship and local innovation) (Gardetti, 2018). Legrand (2020, p. 65) maintains that "sustainability becomes the *sine qua non* of luxury tourism in the near, post-corona future," highlighting the recommendation of a "three-pronged approach to enacting sustainable tourism based on (1) self-imposed assessments; (2) strict regulatory framework; and (3) luxury tourism concept review" (Legrand, 2020, p. 77).

The main objective of this study is to present the first conceptual model applied to sustainable luxury tourism as an innovative theoretical framework, in a pilot version, with the intention to address and manage future challenges related to the sustainability paradigm. It also explores dilemmas in the field of luxury tourism through a set of critical elements that make up the innovative model. This exploratory qualitative research aims to identify the sustainability drivers that are emerging as the most significant triggers for luxury tourism.

This chapter is structured in four parts. First, we present a literature review of the main constructs. After accounting for methodological considerations, we describe and present the development of an assessment framework entitled the "Innovative Model for Sustainable Luxury Tourism." Finally, we conclude with practical implications for managers and recommendations and suggestions for future research.

Literature Review

Luxury Tourism Markets

The Luxury Travel Market (2021), based on the Global Opportunity Analysis and Industry Forecast, delivered a statistical forecast of the Luxury Travel Market between 2021 and 2027.

The Luxury Travel Market was grouped into four key market segments:

1. Age group: Millennial, Generation X, Baby Boomers, and Silver Hair.
2. Type of traveler: absolute luxury, aspiring luxury, and accessible luxury.
3. Type of tour: customized & private vacations, adventure & safari, cruise/ship expedition, small group journey, celebration & special events, and culinary travel & shopping.
4. Region/Geography: North America (USA, Canada, and Mexico); Europe (Germany, France, Spain, Italy, and the rest of Europe); Asia-Pacific (China, India, Singapore, and the rest of Asia-Pacific); and Lamea (Middle East, Latin America, and Africa).

The forecast concludes that:

> The global luxury travel market size was valued at $945.6 billion in 2019 and is expected to grow at a Compound Annual Growth Rate (CAGR) of 11.1% during 2021–2027 to reach 1,198.3 billion in 2027. Luxury travel services cover most desirable and premium experience in terms of luxurious accommodations, convenient transport facilities, and authentic travel experience. The service providers aim to provide outstanding services to its travelers. (Luxury Travel Market, 2021)

Based on these assumptions, the worldwide luxury travel market is forecast to be the fastest-growing segment between 2021 and 2027 (Luxury Travel Market, 2021). The main factors contributing to this growing trend are the increase in middle-class and upper-class disposable incomes, interest in unique and exotic holiday experiences, as well as the growth in the need and desire of people to spend more time with their families. This trend is attributed to the growth of the middle-class population in countries such as China, India, and Malaysia. Despite developed countries such as the USA presenting the largest revenue share, there has been an increase in demand in developing countries such as China and India due to increasing per capita income (Luxury Travel Market, 2021).

Regarding the types of luxury travel—(see Fig. 4.1) classified into customized and private vacations, adventure and safari, cruise/ship expedition, small group journeys, celebration and special events, and culinary travel and shopping—it is possible to infer that the segment with the greatest dominance during the forecasted period is adventure and safari. Conversely, customized and private vacations, cruise/ship expedition, and celebration and special events are the least relevant segments.

Concerning age group (Fig. 4.2), composed of Millennial, Generation X, Baby Boomer, and Silver Hair, all segments are projected to grow but the Millennial segment is forecasted to grow at a maximum CAGR of 11.9%. Generation X and Baby Boomer are two segments that are expected to have a large positive impact; they are expected to present the largest increase compared to Silver Hair, which will only increase slightly.

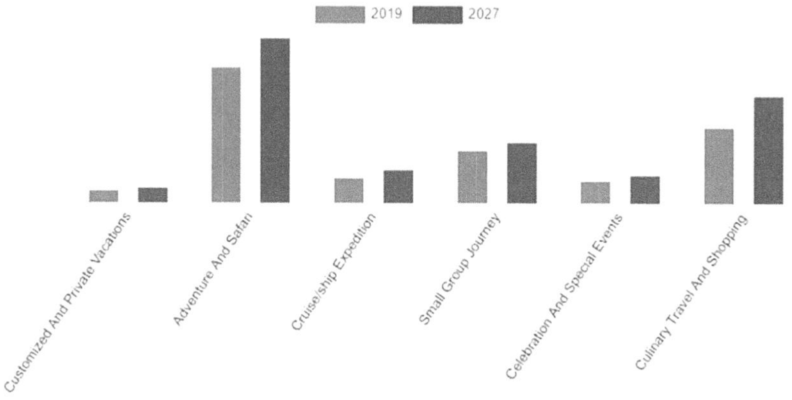

Fig. 4.1 Types of luxury travel tours (Source: Luxury Travel Market (2021))

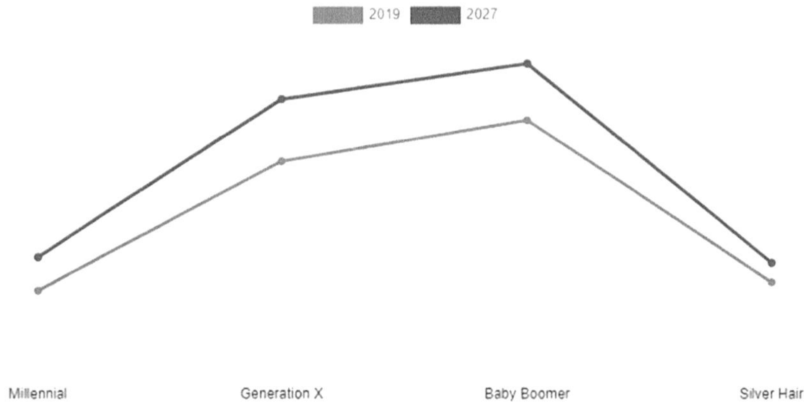

Fig. 4.2 Luxury travelers by age group

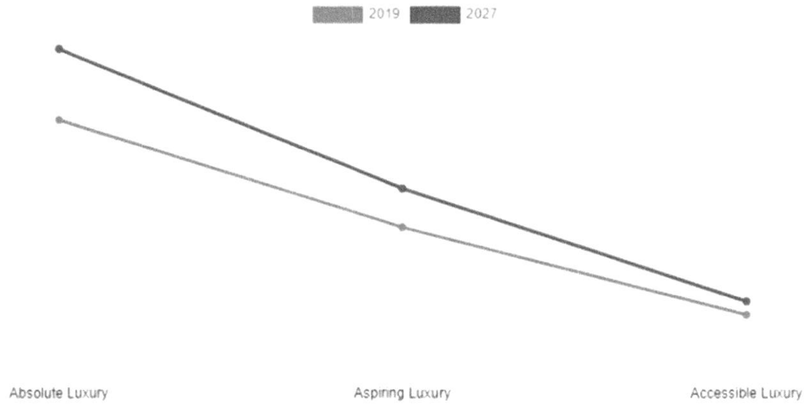

Fig. 4.3 Types of luxury travelers

Figure 4.3 reflects the types of luxury travelers: absolute luxury, aspiring luxury, and accessible luxury. Figure 4.3 indicates that the absolute luxury segment is expected to grow at a maximum CAGR of 12.6%. On the other hand, weak growth is expected in the accessible luxury segment.

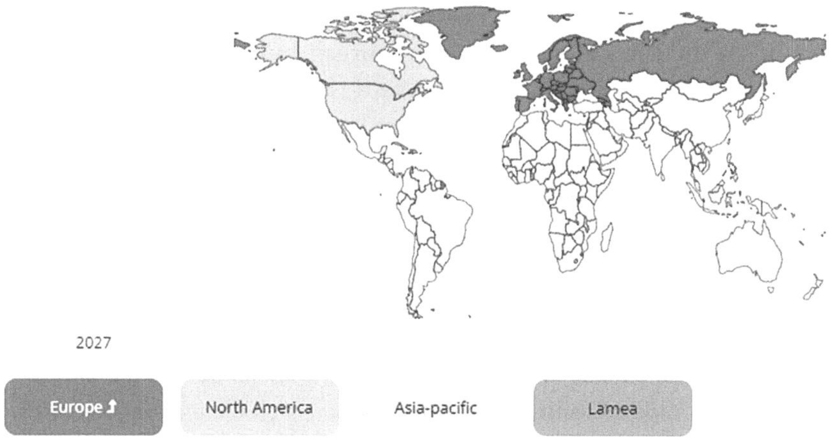

Fig. 4.4 Types of luxury travel by region (Source: Luxury Travel Market (2021))

Figure 4.4 sets out the Luxury Travel Types by region: Europe, North America, Asia-Pacific, and Lamea. Europe is forecasted to reach a CAGR of 9.3% and is set to dominate the market. North America follows it as the second region with the greatest growth potential.

Sustainability in Luxury Tourism

The concept of sustainable luxury is also labeled with other classifications, including deeper luxury (i.e., equivalent to the original concept of traditional luxury tourism) (Bendell & Kleanthous, 2007) and green luxury (Weinstein, 2010). Accompanying this are the growing and emerging environmental and social concerns advocated for by luxury consumers in this niche tourism market (Bendell & Kleanthous, 2007). The question of the compatibility between luxury and sustainability is thus an important one for many tourism destinations (Moscardo & Benckendorff, 2010).

Sustainable luxury tourism is a way to care for, cherish, and maintain destinations (Walker & Moscardo, 2016). Despite the increase in sustainable luxury tourism interest, current theory and practice are underdeveloped. It is widely known and recognized that typical mass tourism is harmful to the environment. As a result, environmental education has taken on a new role for luxury tourism businesses (Jones, 2020). Consequently, Poelina and Nordensvard (2018, p. 149) argue that the "management and protection of landscapes and eco-systems are integral elements of a new wave of sustainable luxury tourism." Developing this further, the authors state that, more than desirable, sustainable luxury tourism is necessary (2019). However, very little is known to date about how sustainability in the luxury tourism sector should work in terms of the value chain concerning the functional and operational performance of the tourism industry.

Against the backdrop of recessions, growing social inequality, and economic crises, sustainability issues have become a challenge for the continued success and growth of the luxury sector. They are considered both a paradox and a provocation by some critics (Kapferer & Michaut, 2015). As a result of new sustainable luxury tourism concepts and ventures, eco-tourism hotels and resorts have emerged. Conventional luxury hotel chains, groups, and brands have advanced toward obtaining sustainability programs and certifications (Jones, 2020).

Table 4.1 sums up the key global problems regarding sustainability and the common characteristics relating to the luxury tourism sought by luxury tourism consumers.

Methodology

Methodologically, we adopt a mixed-method approach based on Porter's Diamond Model (PDM), the Blue Ocean Strategy (BOS), and the Drivers-Pressures-State-Impacts-Responses (DPSIR) framework to achieve greater robustness in order to analyze the credibility criteria used. In addition, we adopt a systematic mapping study (SMS) as the selected research methodological approach to analyze and structure a broad research field concerning methods, designs, and research. We focus on papers relating PDM, BOS, and DPSIR studies published in reliable academic databases such as Emerald, ProQuest, Sage, Science Direct, Scopus, Wiley, Web of Science, Taylor and Francis, and Wiley.

Table 4.1 Global sustainability—key problems in luxury tourism

Problem	Related human actions	Common luxury tourism characteristics
Managing resource use – Rapid decline in non-renewable energy sources – Excessive and inefficient use of freshwater – Use of arable land for other forms of development	– Continued reliance on non-renewable energy in general and in particular for comfort and convenience – Excessive use of freshwater for comfort and convenience – Development of arable land for residential and other commercial activities inefficient use of space	– High usage of energy for features such as air-conditioning and refrigeration of luxury food and beverage – High levels of freshwater consumption in swimming pools, baths, and showers – Large spaces for both individual accommodations and whole properties to provide privacy and exclusive access
Environmental problems – Carbon gas emissions and climate change – Water, air, and ground pollution – Waste disposal – Habitat destruction and biodiversity loss	– Use of inefficient and carbon-based fuels Pollution and waste generated from production and usage associated with extensive consumption of non-essential goods and services – Destruction of habitat to support production	– Long distance travel to reach many locations involving large carbon emissions – Habitat damage for building – Provision of seafood in dining – Larger waste issues associated with more intensive consumption – Chemicals associated with cleaning and cosmetics provided

(*continued*)

Table 4.1 (continued)

Problem	Related human actions	Common luxury tourism characteristics
Equity in socio-economic development – Exploitation of lower standards of living to support cheaper production of goods and services for consumption elsewhere – Poor working conditions include lower pay and limited training – Economic leakage from foreign-owned businesses – Limited distribution of economic activities to local community – Loss of local culture – Asymmetry in ownership of key resources increasing local dependency and vulnerability	– Globalization of the economy where cheaper production costs in the global South allow for increased consumption in the global North – Reliance on post-colonial models of development	– Often provide luxury tourism opportunities in countries with lower standards of living because of cheaper wages and land prices – Multinational/foreign ownership, with need to import goods to support luxury tourist expectations and to use non-local staff and managers to meet international luxury tourism standards, support high economic leakage and limited benefits for local communities – Destruction or commodification of local culture

We chose this mixed approach because all four approaches address business strategy and competitiveness in their model. They also share dimensions, factors, elements, and characteristics in their basic matrices. We use these interrelated methods as they emerge as the most suitable methods to advance the main strategic and emerging issues, trends, and challenges related to the whole competitive advantage of the sustainable luxury tourism industry.

This methodology integrates a set of studies in terms of themes; hence, we are in a better position to describe current knowledge and practice, evaluate progress, identify gaps, and plot a course for future research (Valeri & Baggio, 2020c). The literature review has long been described as "a vital and popular technique in the consumer researcher's toolkit" (Valeri & Baggio, 2020c).

PDM: Porter's Diamond Model

The Diamond Model proposed by Porter (1990) determines the competitive advantage and success of companies, firms, or sectors within national boundaries. Porter's Diamond Model comprises four determinant factors in the competitive advantage of a nation, region, or cluster, namely, (1) factor conditions: the endowment of a country with factors of production such as the specialist labor skills or infrastructures necessary to the ongoing competitive activities of a specific industry; (2) strategic structure and business rivalries: conditions that nationally regulate the founding, organization, and management of firms and companies and the nature of their internal rivalries; (3) demand conditions: the characteristics of the internal demand for a specific good or service, especially the presence of sophisticated and demanding clients; (4) existence of related and support industries: the existence of supply sectors and/or related chains that attain international market competitive standards (Porter, 1990).

Some authors (e.g., Ozer et al., 2012; Estêvão et al., 2018) have applied Porter's Diamond Model to the field of tourism. In particular, Adillón (2019) adopted the Diamond Model in the context of sustainable tourism, determining seven elements to ensure the adequate development of sustainable tourism activities. These are the carrying capacity, local community, tourist experience, destination management organization, seasonality management, demand management, and smart destinations. Furthermore, the Diamond Model was also the selected method in a study of a coastal tourism commodity industry cluster (Luo, 2019).

BOS: Blue Ocean Strategy

Kim and Mauborgne (2005) pioneered the Blue Ocean Strategy (BOS) concept. The market consists of two types of oceans: the "Red" and the "Blue." The Red Ocean refers to the low-profit saturated market with several competitors. In contrast, the Blue Ocean has a lesser-known market and almost no competition, making it less exploited by companies. The purpose of the Blue Ocean is to describe the breadth and depth of the potential market, alongside the unknown or unexplored exploitation. The BOS includes a framework of four actions: eliminate, reduce, create, and raise. The five main traits of BOS are as follows: (1) create uncontested market space; (2) make the competition irrelevant; (3) create and capture new demand; (4) break the value-cost trade-off; and (5) align the whole system of the activities of a firm with its strategic choice of differentiation and low cost (Kim & Mauborgne, 2005).

The BOS has been conceptualized and developed as an innovative market strategy for tourism. It has been used in the hotel industry (Ahmat et al., 2015) in order to allow hoteliers to summarize the BOS attributes related to the applied strategy for long-term sustainability. It has also been applied in the wine tourism industry (Priilaid, 2019) and cultural product tourism (Mohamed & Abaza, 2020), highlighting the innovation of some modern technological methods in the context of heritage sites.

DPSIR: Drivers-Pressures-State-Impacts-Respon ses Framework

For Rempel (2012), the Drivers, Pressures, State, Impacts, and Responses (DPSIR) approach has been underexplored in tourism, but it is a viable tool to support a more critical analysis on sustainability in this area. More recently, Moscardo (2017) adapted the DPSIR framework in the context of sustainable luxury in hotels and resorts to contextualize the emergence of sustainable luxury tourism. Another study applied the DPSIR model in the sustainable tourism development context, analyzing the interregional similarities and differences in selected Mediterranean regions (Pivčević et al., 2020). Another perspective, to develop the concept further, is a case study that connected ecotourism with sustainable tourism as having integrated the DPSIR framework (Swangjang & Kornpiphat, 2021).

Results

We developed the four key elements that compound the Innovative Model for sustainable luxury tourism (per Fig. 4.5) by gathering valuable insights from four models and frameworks, namely, Porter's Diamond Model, Blue Ocean Strategy, DPSIR (Drivers, Pressures, State, Impacts, and Responses) framework, and the APIC (Analyze, Planning, Implementation and Control) model.

Figure 4.5 shows the development of the Innovative Model for Sustainable Luxury Tourism as an assessment tool, comprising four key elements: Support Structures & Features, Governance Ecosystem, Accommodation & Services, and Key Stakeholders. Each of the four elements is subdivided into a total of twenty topics.

The model consists of an innovative conceptual model applied to sustainable luxury tourism as an innovative theoretical framework addressing and managing future challenges related to the sustainability paradigm

Fig. 4.5 Innovative model for sustainable luxury tourism

and dilemma in the field of luxury tourism. It is characterized as a hybrid and systematic model that simultaneously correlates the four elements (Support Structures & Features; Governance Ecosystem; Accommodation & Services; and Key Stakeholders), as a result of its conceptual, strategic, tactical, and operational gearing.

This novel taxonomy serves as a useful tool for systematic, integrative, and synergic best benchmarking practices underlying sustainable luxury tourism in addition to its boundaries, features, and key components.

The examination of theoretical and practical implications of this framework methodology contributes to the development of scientific knowledge about the application of the concepts for the Innovative Model for Sustainable Luxury Tourism through the development of an assessment framework (Valeri, 2016; Valeri & Baggio, 2020c).

Conclusion and Implications

The Innovative Model for Sustainable Luxury Tourism demonstrates that it is possible to produce and deliver luxury tourism and hotel goods and services more sustainably. In conclusion, the Innovative Model for Sustainable Luxury Tourism has implications with regard to policy-making or future business development plans, such as developing sustainable and ethical marketing strategies and plans (with the local community), promoting smart luxury tourism destinations, and ensuring the compliance of ESG (Environmental, Social, and Governance) criteria, regarding competitive differentiation and advantage.

The sustainable luxury tourism of the future is one that provides total smart and green luxury comfort, is environmentally conscious, preserves natural resources, and is supported by future tourist generations concerned with environmental issues and practices.

Indeed, in the future, sustainability will play a significant role in the new era of luxury tourism. In its pilot version, the proposed innovative theoretical framework for sustainable luxury tourism can be considered the missing trigger, challenge, and opportunity for sustainable luxury tourism to develop its nature management and competitive governance model. Therefore, luxury tourism through sustainability could very well be the disruptive innovation needed to counter unsustainable luxury hotel consumption observed in hotels and resorts, among others. The most important aspect to highlight is that sustainable luxury tourism will work best if it is achieved by focusing on the elements that make up the diamond model, which aims at a new era of tourists.

Regarding limitations, it was noted that there are still a significant number of ongoing tensions between the concepts of sustainability and luxury that create some barriers and difficulties for the management of this emerging form of tourism worldwide. These limitations prevent a more complete and in-depth analysis of the current and future underlying determinants. Much of this is due to tensions derived from the COVID-19 pandemic, which highly conditioned both governance and stakeholders.

As for the practical implications for management, sustainability agendas in luxury tourism must address local realities and communities. Sustainability agendas in luxury tourism rely on local infrastructure and have to respond to local stakeholder expectations, including standardized requirements and global, regional, local, and individualized sustainability action plans. Exploring creative ideas, novel solutions, and innovative strategies are crucial in this industry when facing increased environmental and societal challenges, especially considering the size and structure of the tourism and hospitality sectors.

For a future sustainable development agenda, sustainable luxury tourism should be valued option for government, stakeholders, shareholders, tourists, tour operators, and travel agents, resulting in a focus on sustainable development goals. Regarding future research, we plan to advance the assessment framework developed in this exploratory qualitative study and to apply the Innovative Model for Sustainable Luxury Tourism to a set of key strategic stakeholders in order to analyze and evaluate the model in the practical and operational context of sustainability in luxury hotels and resorts.

References

Adillón, R. (2019). DIAMOND MODEL: A theoretical framework for the sustainable development of tourism. *Arts & Humanities Open Access Journal, 3*(1), 11–23. https://doi.org/10.15406/ahoaj.2019.03.00099

Ahmat, N. C., Abas, R., Ahmad-Ridzuan, A. H., Radzi, S. M., & Zahari, M. S. M. (2015). Blue Ocean strategies in hotel industry. In S. M. Radzi, N. Sumarjan, C. T. Chik, M. S. M. Zahari, Z. Mohi, & M. F. S. Bakhtiar (Eds.), *Theory and practice in hospitality and tourism research* (pp. 43–47). CRC Press.

Amatulli, C., De Angelis, M., & Stoppanib, A. (2021). The appeal of sustainability in luxury hospitality: An investigation on the role of perceived integrity. *Tourism Management, 83*, 1–12. https://doi.org/10.1016/j.tourman.2020.104228

Bendell, J., & Kleanthous, A. (2007). *Deeper luxury*. WWF-UK.

Correia, A., Kozak, M., & Del Chiappa, G. (2020). Examining the meaning of luxury in tourism: A mixed-method approach. *Current Issues in Tourism, 23*(8), 952–970. https://doi.org/10.1080/13683500.2019.1574290

Estêvão, C., Nunes, S., Ferreira, J., & Fernandes, C. (2018). Tourism sector competitiveness in Portugal: Applying Porter's Diamond. *Tourism & Management Studies, 14*(1), 30–44. https://doi.org/10.18089/tms.2018.14103

Esty, D. C., & Winston, A. S. (2009). *Green to gold*. John Wiley & Sons.

Gardetti, M. A. (2018). Entrepreneurship, innovation and luxury: The case of ANTHYIA. In M. A. Gardetti & S. S. Muthu (Eds.), *Sustainable luxury, entrepreneurship, and innovation* (pp. 57–80). Springer. https://doi.org/10.1007/978-981-10-6716-7

Jones, G. (2020). Luxury tourism and environmentalism. In *The Oxford handbook of luxury business*. Oxford University Press. https://doi.org/10.1093/oxfordhb/9780190932220.013.31

Kapferer, J.-N., & Michaut, A. (2015). Luxury and sustainability: A common future? The match depends on how consumers define luxury. *Luxury Research Journal, 1*(1), 3–17. https://doi.org/10.1504/LRJ.2015.069828

Kim, W. C., & Mauborgne, R. (2005). Blue Ocean strategy: From theory to practice. *California Management Review, 47*(3), 105–121.

Lansing, P., & De Vries, P. (2007). Sustainable tourism: Ethical alternative or marketing ploy? *Journal of Business Ethics, 72*(1), 7–85.

Legrand, W. (2020). Luxury tourism in remote destinations: An inquiry into sustainability. In J. S. Chen (Ed.), *Advances in hospitality and leisure* (Vol. 16, pp. 65–81). Emerald Publishing Limited. https://doi.org/10.1108/S1745-354220200000016006

Luo, X.-Y. (2019). Coastal tourism commodity industry cluster based on diamond model and ecological niche. In D. Gong, H. Zhu, and R. Liu (Eds.), *Selected topics in coastal research: Engineering, industry, economy, and sustainable development. Journal of Coastal Research, 94*, 828–832.

Luxury Travel Market. (2021). Allied Market research. https://www.alliedmarketresearch.com/luxury-travel-market

Machado, A. F., & Sousa, B. (2018). Sustainable luxury in the contexts of hospitality and tourism: From the competitive differential to the concern with social responsibility. *International Journal of Marketing, Communication and New Media, 4*, 28–56.

Mohamed, G., & Abaza, W. (2020). Applying Blue Ocean strategy to culture tourism product: The case of Egypt. *Global Review of Research in Tourism, Hospitality and Leisure Management, 6*(1), 776–788.

Moscardo, G. (2012). Building social capital to enhance the quality-of-life of destination residents. In M. Uysal, R. Perdue, & M. J. Sirgy (Eds.), *Handbook of tourism and quality-of-life research* (pp. 403–422). Springer.

Moscardo, G. (2017). Sustainable luxury in hotels and resorts: Is it possible? In M. Gardetti (Ed.), *Sustainable management of luxury. Environmental footprints and eco-design of products and processes* (pp. 163–189). Springer. https://doi.org/10.1007/978-981-10-2917-2_8

Moscardo, G., & Benckendorff, P. (2010). Sustainable luxury: Oxymoron or comfortable bedfellows? In M. van Niekerk, W. Coetzee, & S. Geldenhuys (Eds.), *Proceedings of global sustainable tourism conference. Global sustainable tourism* (pp. 709–728). Mbombela.

Ozer, K., Latif, H., Sarusik, M., & Ergun, O. (2012). International competitive advantage of Turkish tourism industry: A comperative analyse of Turkey and Spain by using the diamond model of M. Porter. *Procedia—Social and Behavioral Sciences, 58*, 1064–1076.

Pivčević, S., Petrić, L., & Mandić, A. (2020). Sustainability of tourism development in the Mediterranean—Interregional similarities and differences. *Sustainability, 12*, 7641. https://doi.org/10.3390/su12187641

Poelina, A., & Nordensvard, J. (2018). Sustainable luxury tourism, Indigenous communities and governance. In M. Gardetti & S. Muthu (Eds.), *Sustainable luxury, entrepreneurship, and innovation* (pp. 147–166). Springer. https://doi.org/10.1007/978-981-10-6716-7_8

Porter, M. (1990). *The competitive advantage of nations*. Free Pass.
Priilaid, D. (2019). Exploring Blue Ocean innovation in the wine industry. In M. Sigala & R. Robinson (Eds.), *Management and marketing of wine tourism business* (pp. 241–259). Palgrave Macmillan. https://doi.org/10.1007/978-3-319-75462-8_12
Rempel, J. M. (2012). Exploring the causal nexus of tourism impacts on quality-of-life. In M. Uysal, R. Perdue, & M. J. Sirgy (Eds.), *Handbook of tourism and quality-of-life research* (pp. 583–606). Springer.
Ronchi, E., Alberto, F., & Musmeci, F. (2002). A system oriented integrated indicator for sustainable development in Italy. *Ecological Indicators, 2*(1–2), 197–210. https://doi.org/10.1016/S1470-160X(02)00045-6
Swangjang, K., & Kornpiphat, P. (2021). Does ecotourism in a Mangrove area at Klong Kone, Thailand, conform to sustainable tourism? A case study using SWOT and DPSIR. *Environment, Development and Sustainability, 23*, 15960–15985. https://doi.org/10.1007/s10668-021-01313-3
Valeri, M. (2016). Networking and cooperation practices in the Italian tourism business. *Journal of Tourism, Heritage and Services Marketing, 2*(1), 30–35. https://doi.org/10.5281/zenodo.376333
Valeri, M., & Baggio, R. (2020c). A critical reflection on the adoption of blockchain in tourism. *Journal Information Technology and Tourism, 23*(2), 121–132. https://doi.org/10.1007/s40558-020-00183-1
Walker, K., & Moscardo, G. (2016). Moving beyond sense of place to care of place: The role of Indigenous values and interpretation in promoting transformative change in tourists' place images and personal values. *Journal of Sustainable Tourism, 24*(89), 1243–1261. https://doi.org/10.1080/09669582.2016.1177064
Weinstein, J. (2010). Green luxury no oxymoron. *Hotels, 44*(2), 30.

Part III

Sustainable Tourism Innovations and Solutions

5

Recreation in Forests: Implications from the COVID-19 Pandemic

Patricia Poehler and Monika Bachinger

Introduction

Forests contribute to human wellbeing in many ways. They provide food and drinking water and contribute to climate regulation and soil formation. This chapter focuses on the cultural ecosystem services (CES) of forests. CES are intangible services that are co-created by humans and forests (Fish et al., 2016b). Examples of CES are recreational, aesthetic, creative, and learning opportunities (MA, 2005).

In principle, all types of nature may be adequate for CES. However, forests exhibit a three-dimensional structure and are composed of different layers (from the humus layer to the treetop) and therefore offer a unique visual appearance. The forest climate also differs from open areas as it offers a cooler and more humid environment in warmer

---

P. Poehler • M. Bachinger (✉)
University of Applied Forest Sciences, Rottenburg, Germany
e-mail: Patricia.Poehler@lrasig.de; bachinger@hs-rottenburg.de

T. Walker et al. (eds.), *Sustainable Tourism*,
https://doi.org/10.1007/978-3-031-43528-7_5

temperatures (O'Brien et al., 2017). These characteristics encourage people to become physically active within such spaces (Hartig et al., 2011). These spaces also offer a place to meet and interact with others and therefore perform a social and integrative function (Jay & Schraml, 2009). Most importantly, forests account for positive psychological impacts. After forest visits, people report feeling refreshed and energetic and notice a reduction in stress and depression (Lee et al., 2011; Korpela et al., 2014).

These benefits of forests became increasingly important during the COVID-19 pandemic. In March of 2020, the World Health Organization declared a pandemic emergency, which led to the closure of schools, businesses, and recreation facilities worldwide (Ferguson et al., 2022). Thus, common leisure activities like going to a gym, meeting with friends in a restaurant, or watching a movie became impossible. The only recreation spaces that could be visited were outdoors, especially green spaces, such as forests. The importance of forests for recreation therefore increased (Bamwesigye et al., 2021; Derks et al., 2020; Grima et al., 2020; Weinbrenner et al., 2021).

However, the extent to which forests actually contributed to human wellbeing during the pandemic has not yet been considered in detail with the exception of Muro et al. (2022), who focus on mindfulness practices (forest bathing) during COVID-19. Although it is known that forests positively impact human wellbeing, it is unclear how strong these impacts were during COVID-19. Thus Bamwesigye et al. (2021) have called for research on the relationship between forest ecosystem services and COVID-19 stresses and mental health. Against this background, this chapter looks at how and why visits to forests changed during the COVID-19 pandemic and to what extent forests contributed to human wellbeing in the area of Baden-Württemberg, Germany during this time. Based on the results of our case study and the observation that forest administrations faced challenges in visitor management during the pandemic, we formulate recommendations for the management of forests as recreational areas in a post-COVID era.

Literature Review

Importance of Forests for Human Wellbeing

Forests contribute to human wellbeing in different ways. Forests protect against erosion, contribute to global climate regulation, and retain water and improve landscape aesthetics (EEA, 2016). Forests provide habitat for a large number of animals and contain a significant proportion of plant species (EEA, 2016). In addition, forests are economically relevant, especially due to the commercial use of wood (Beland Lindahl & Westholm, 2011). The focus of this chapter, however, is on the contribution of forests to recreation and wellbeing. Forests play a particularly important role in this field: "Forests attract a large number of visitors who appreciate nature, biodiversity and peaceful surroundings" (EEA, 2016). The demand for recreation in forests is projected to increase because of urbanization and population growth (Lupp et al., 2016).

Forest visitors profit from a wide array of positive effects on wellbeing: "Recreation contributes greatly to the physical, mental and spiritual health of individuals, families and friends, and instills pride in natural and cultural heritage" (White et al., 2016). In general, existing research acknowledges that forests provide space for sporting activities that strengthen physical wellbeing. Forests differ from man-made environments in their visual structure, colors, and soundscape and therefore help visitors relax (Brown et al., 2013). Forests, with their three-dimensional structure, disperse great numbers of visitors, therefore mitigating the effects of overcrowding which can occur in man-made spaces. Lastly, forested spaces are not expensive to visit compared to other recreational facilities, thus helping to blur social differences (O'Brien & Morris, 2014).

In recent years, recreational behavior in forests has diversified and intensified. Typical recreational activities in forest are hiking/walking (either alone or with pets), biking, and playing with children (UK Forest Research, 2021). Other activities have enjoyed increased popularity recently, such as visiting nature trails, adventure experiences, or geocaching. Overall numbers of forest visits are projected to continue growing (Lupp et al., 2016). The degree of change is dependent on factors like the increase of population in the surrounding area, but also age, gender, ethnicity, and available income (White et al., 2016). Visits to forests depend on certain socio-economic factors and are not evenly spread over society (O'Brien & Morris, 2014).

Wellbeing as Cultural Ecosystem Service

Cultural ecosystem services (CES) are often defined as "non-material benefits people obtain from ecosystems through spiritual enrichment, cognitive development, reflection, recreation and aesthetic experience" (MA, 2005). Examples of CES are nature-based opportunities for tourism and recreational activities, inspiration for art and design, place attachment, spiritual and religious inspiration, as well as education and training. In particular, Dickinson and Hobbs (2017) cite outdoor recreation and related health effects as examples of CES.

In recent years, scholars have increasingly acknowledged the importance of CES for human wellbeing: "CES provide an important link between the biophysical environment and human well-being" (Bryce et al., 2016). According to Pleasant et al. (2014), CES influence all four categories of human wellbeing, that is safety, health, social inclusion, and basic needs such as nutrition. Furthermore, CES contribute to sense of place, identity, and autonomy which are all of great importance for human wellbeing (Bryce et al., 2016; Stålhammar & Pedersen, 2017).

CES, however, are not ready-to-use services (Fischer & Eastwood, 2016; Fish et al., 2016a). They have to be understood as relational constructs and be actively "co-produced" by human beings interacting with nature (Fish et al., 2016b). Therefore, when examining CES, it is crucial to consider both the characteristics of green spaces and the social and cultural context in which they emerge (Dickinson & Hobbs, 2017; Hegetschweiler et al., 2017). The characteristics of green spaces, including their size or their physical and infrastructural features might influence what kind of CES emerge (Hegetschweiler et al., 2017). On the other hand, forest visitors and their values, experiences, and recreational behavior influence the CES as well. Other factors which determine the generation of CES are educational background, age, gender, and cultural background of visitors (Dickinson & Hobbs, 2017). Additionally, cultural practices like how the space is used (i.e., hiking versus hunting) influence CES (Fish et al., 2016b).

Changes in Forest Recreation due to COVID-19

During the pandemic, people increasingly spent time at home. Contact with friends was reduced. Working from home presented challenges that put a strain on wellbeing. In Germany, for example, a survey found that 47% of participants felt that the pandemic had changed their lives either heavily or very heavily (DW, 2021). Worries about livelihood increased, as did feelings of sadness and loneliness (Buecker et al., 2020). However, fewer people reported suffering from stress or exhaustion than before the pandemic (Ohlbrecht & Jellen, 2021). With regard to the question of whether overall subjective wellbeing had changed, studies come to different conclusions. While Entringer et al. (2020) report a rather unchanged life satisfaction compared to the previous year, Ohlbrecht and Jellen (2021) and Ravens-Sieberer et al. (2022) indicate that general life satisfaction decreased.

Forests offered one of the few recreation opportunities during the periods of lockdown in the region. This particularly affected forests near cities. In a forest area near Bonn, for example, more than twice as many visitors were counted during the first lockdown in 2020 than in previous years. While previously there were clear peaks in visitation in the morning and evening hours, visitors were now distributed more evenly throughout the day. In addition, visitation rates on weekdays hardly differed from the weekend (Derks et al., 2020). A US-based study came to similar conclusions, showing that around 69% of 400 respondents visited a city forest in Burlington (Vermont) more frequently during the pandemic. Many of the visitors had never or rarely been in forests before (Grima et al., 2020). This applies to Germany as well, where inexperienced forest visitors accounted for a larger percentage of the total population than is observable in pre-pandemic figures (Derks et al., 2020).

This increase in forest visitation even reached societal groups which are usually underrepresented. In particular, women, the elderly, and sick people, but also socially disadvantaged people or migrants who typically visit green spaces such as forests less frequently (Boyd et al., 2018). The demographic representation of forest visitors evened during the COVID-19 pandemic. In particular, visitors identifying as female became more common (OIA, 2021). On the other hand, various forms of conflict emerged, often between locals and out-of-region visitors. Locals felt deprived of their usual leisure routines in forests by an overwhelming number of visitors (Rice et al., 2020).

Research Design

Study Area

To gain insight into changes in forest visitation during the COVID-19 pandemic and the contribution of forests to wellbeing during that time, we conducted an online survey in the federal state of Baden-Württemberg, Germany. Baden-Württemberg is in the southwest of Germany and has about 11 million inhabitants. With a forest share of 38% of the state area, it is one of the most forested states in Germany. There are around 1.4 million hectares of forest. Even before the pandemic, 66% of the population in Baden-Württemberg visited forests at least once a week. Most inhabitants were able to reach forests in less than 15 minutes. Once there, they stayed for an average of 73 minutes (Ensinger et al., 2012). In general, popular activities include walking (63.7%), sports and health activities (30.2%), "just being" (31.6%), and observing nature (26.5%) (Hunziker et al., 2013). This correlates to studies from other countries (UK Forest Research, 2021). Forests in Baden-Württemberg are visited for a variety of reasons: "getting in fresh air", "being in nature", and "physical exercise". For many visitors, however, a visit to forests simply meant a contrast to everyday life, which allows for relaxation (Ensinger et al., 2012).

Data Collection and Data Analysis

We divided the online survey into four major sections: the first section contained questions on the number, duration, and importance of forest visits before and during the COVID-19 pandemic. In the second section, participants were questioned on the changes in the motives and activities of their forest visits. The third section addressed the respondents' wellbeing during the COVID-19 pandemic, and in the fourth section, questions collected socio-demographic data.

The first section of the survey contained questions which focused on changes in forest visitation due to the pandemic. The related items were:

- Since the beginning of the COVID-19 pandemic I have been in forests more often.
- Since the beginning of the COVID-19 pandemic my visits to forests have been longer.
- Even before the COVID-19 pandemic, visiting forests was an essential part of my leisure activities.

All three questions could be answered on a 5-point Likert scale. Furthermore, two questions with numerical value answers helped to collect data on visitation frequency and length of stay during the pandemic. The questions were:

- During the last two weeks, how often have you been to forests?
- During the last two weeks, how long would you say did your forest visits take on average?

In the second section, we formulated questions in order to help record changes in the motives and activities of the forest visits. Respondents indicated whether they performed activities significantly less or significantly more frequently. As for motives, they indicated which motives (from a given list) had become more important during COVID-19. For both questions, existing survey instruments were used. The motives for forest visits come from Ensinger et al. (2012), and the forest-related activities were based on Hunziker et al. (2013). The related questions were:

- During COVID-19, which of the given motives for forest visits have become more important to you?
- To what extent have you changed to practice the following activities in forests?

The first question was answered by selecting motives from a list, and the second question was answered on a 5-point scale ranging from 1 = much less frequently to 5 = much more frequently.

The third section of the online survey related to wellbeing during the pandemic. The options correspond to the WHO-5 wellbeing questionnaire (WHO, 1998). The related questions were:

- Over the last two weeks

  (a) I have felt cheerful and in good spirits.
  (b) I have felt calm and relaxed.
  (c) I have felt active and vigorous.
  (d) I woke up feeling fresh and rested.
  (e) My daily life has been filled with things that interested me.

Answers could be given on a 6-point scale ranging from 5 = all the time to 0 = at no time. A second question in this section recorded changes in subjective wellbeing before and during the pandemic. The question was:

- Compared to before COVID-19, to what extent has your wellbeing changed?

Answers were given on a 6-point scale from 1 = no at all to 6 = very strong.

The fourth section of the survey aimed at collecting a few relevant socio-demographic data points, such as age, gender, level of education, and housing situation. The questionnaire was deployed online and distributed as widely as possible with the help of social media, email, and the distribution list of the Baden-Wuerttemberg Chamber of Forestry. We evaluated the data obtained using SPSS Statistics 27.

Results

The survey was conducted between April 7 and April 20, 2020. A total of 422 people took part. After excluding cases with a high number of missing values, 345 questionnaires remained. The sample consisted of slightly more men (187) than women (156). Around half of the respondents

(55.4%) were under 30 years old. With regard to their housing situation, the vast majority of respondents reported having sufficient space for retreat, for example, access to a balcony, a terrace, or garden. However, only 44% stated that they had enough space for their children to play; 43.6% live alone or as a couple. All in all, the sample does not represent the average population of Baden-Württemberg. In particular, young individuals are overrepresented. Furthermore, the housing situation does not reflect the average situation in Baden-Wüerttemberg.

Forest Visitation Patterns

These limitations notwithstanding, our results show that, compared to pre-pandemic times, both the frequency and the duration of forest visits changed for a observable part of respondents, though not for the majority. Around 40% agreed to the question that they had been in forests more often since the beginning of the pandemic; 45% disagreed with this question. When asked how often they stayed in forests during the last two weeks, around 82% stated that they had been in forests at least once a week. However, significant percentages of the respondents visited forests more than that—sometimes several times a day. As regards length of each stay, around one-third confirmed that forest visits took longer than in pre-pandemic times. A significant part of forest visits lasted half an hour to an hour (31.7%). However, some people extended their visits by up to three hours. With regard to the importance of forests for leisure, a great majority answered that visiting a forest had already been an important part of leisure activities before the pandemic (see Table 5.1).

Motives and Activities in Forests

Results on forest activities showed that walking (63.9%), jogging (50.0%), and cycling (44.3%) were practiced more frequently during the pandemic. Walking the dog, hiking, and observing nature did not incur a change. Playing with children, instead, was an activity practiced less by 42.7% of respondents during the pandemic. Changes in motives for visiting forests were rather small. First and foremost, there was an increase in the motive of physical activity, to which 20% of responses referred. Other motives with a noticeable increase in importance were "being outside" with 13.50% and "being in nature" with 11.76%. Other frequently mentioned motives of forest visits, which gained in importance during the pandemic were "taking in fresh air", "being able to switch off", "having a rest", and "being away from everyday life".

Table 5.1 Changes in the frequency, length, and importance of forest visits during the pandemic (percentage of responses)

	Fully disagree (1)	Disagree (2)	Neither agree nor disagree (3)	Agree (4)	Strongly agree (5)	Mean	Std. dev.
	Percentage						
Since the beginning of the COVID-19 pandemic my visits to forests have been longer (*n* = 339)	26.5	21.5	18.3	19.5	14.2	2.73	1.40
Since the beginning of the COVID-19 pandemic I have been in forests more often (*n* = 340)	21.5	23.5	16.2	24.7	14.1	2.86	1.38
Even before the COVID-19 pandemic, visiting forests was an essential part of my leisure activities (*n* = 343)	4.7	12.8	22.2	23.0	37.3	3.76	1.21

Source: Own illustration

Subjective Wellbeing

When asked about changes in their subjective wellbeing compared to pre-pandemic times, almost 60% of respondents stated that their wellbeing had changed "not at all" or "only slightly." Around 40% indicated a strong to very strong change. The mean is 3.32. The direction of the change, however, was not recorded and thus could be negative as well as positive. With regard to the current state of wellbeing, the respondents answered to items that related to their condition in the past two weeks (see Table 5.2). Results showed that the respondents mostly felt cheerful and in good spirits, calm, and relaxed (mean = 3.10; 3.01). The item agreed to the least was that their daily life was mostly full of things that interested them (mean = 2.62). The mean value of the sum score of the five items is 13.99 (standard deviation 4.92, minimum 0, maximum 25).

Table 5.2 Subjective wellbeing (percentage of answers)

Over the last two weeks	At no time (0)	Some of the time (1)	Less than half of the time (2)	More than half of the time (3)	Most of the time (4)	All the time (5)	Mean	Std. dev.
	Percentage							
I have felt cheerful and in good spirits	0.9	5.3	40.9	26.9	13.2	12.9	3.10	1.157
I have felt calm and relaxed	1.5	6.2	35.9	27.9	14.4	14.1	3.01	1.205
I have felt active and vigorous	2.4	4.1	25.1	26.0	20.9	21.5	2.62	1.249
I woke up feeling fresh and rested	8.5	7.9	23.8	26.7	15.8	17.3	2.64	1.430
My daily life has been filled with things that interested me	3.5	6.7	22.9	26.1	18.2	22.6	2.62	1.328

Source: Own illustration

Influence of Forest Visitation on Wellbeing

The next step of our analysis looked at the influence of forest visits on subjective wellbeing. We performed a regression analysis to test whether longer and more frequent visits to forests had a positive effect on the subjective wellbeing of respondents. Secondly, we tested whether longer and more frequent forest visits influenced changes in subjective wellbeing of respondents compared to pre-pandemic times. The results, presented in Table 5.3, show that longer visits to the forest have a positive effect on current wellbeing, albeit to a small extent ($\beta = 0.196$, $p = 0.036$). More frequent forest visits showed a positive influence on changes in wellbeing compared to pre-pandemic times ($\beta = 0.272$, $p = 0.003$) (Table 5.3).

Discussion

We discuss the results of this study by relating them back to findings from literature. The main findings are summarized in Table 5.4. Based on the results detailed in the previous section, we conclude an increase in the frequency of forest visits. Compared to the study by Ensinger et al. (2012), which was done in pre-pandemic times, the frequency of visits to the forest increased by 16%. Compaired with approximately 66% in pre-pandemic times, now 82% of respondents visited forests at least once a week. This agrees with the observations of Derks et al. (2020) and Grima et al. (2020), who both found an increase in the number as well as the frequency of forest visits during the COVID-19 pandemic. However, in this sample it is unclear whether the change was induced by the pandemic. Most respondents to the study did not report a substantial increase in forest visits since the beginning of COVID-19. Instead it might be assumed that they practiced frequent forest visits before the pandemic already.

With regard to the length of stay, there are small differences in this study compared to findings from literature conducted prior to the COVID-19 pandemic (Ensinger et al., 2012; Hunziker et al., 2013; UK Forest Research, 2021). Most visitors (31.7%) stayed in forests 30 to

Table 5.3 Influence of length and frequency of forest visits on subjective wellbeing

	Influence of forest visits on wellbeing over the last two weeks				Influence of forest visits on changes of wellbeing compared to pre-pandemic time			
	Unstandardized coefficients		Standardized coefficients		Unstandardized coefficients		Standardized coefficients	
Variable	*B*	Std. error	Beta	Sig.	*B*	Std. error	Beta	Sig.
Since the beginning of the COVID-19 pandemic my visits to forests have been longer	0.687	0.326	**0.196**	0.036	-0.141	0.078	-0.167	0.071
Since the beginning of the COVID-19 pandemic I have been in forests more often	-0.200	0.333	-0.056	0.548	0.235	0.080	**0.272**	0.003
Model								
R Square	0.024				0.028			
Adjusted *R* Square	0.018				0.023			
F	3.966				4.873			
Sig.	0.020				0.008			

Source: Own representation

Table 5.4 Overview on main findings on forest visitation during the COVID-19 pandemic

	Findings from the study
Frequency of forest visits	Respondents visited forests more often than another sample in pre-pandemic times; however, influence of the pandemic on visitation pattern is unclear.
Length of stay	Respondents stayed a little bit longer in forests than another sample before the pandemic; however, influence of the pandemic on visitation pattern is unclear.
Visitation motives and activities	The overall structure of forest visitation motives and activities changed slightly
Subjective wellbeing	Respondents as frequent forest visitors expressd a positive effect of forest visits on wellbeing during the pandemic
Influence of length of stay on wellbeing	Longer stays in forests positively affected perceived wellbeing, limitations see text
Influence of frequency of visits on wellbeing	More frequent visits impacted on the change of wellbeing during COVID-19, limitations in text
Key finding and link to managerial implications	**Forests make important contributions to human wellbeing; they need to be managed accordingly; recreational issues need to be considered more comprehensively in forest management. Forest management strategies, competencies, and measures have to be adapted.**

60 minutes. However, more than half of the respondents stayed longer, even up to more than three hours. Data suggests that the average length of stay was between 60 and 90 minutes. This is slightly more than reported by Ensinger et al. (2012), who found that most visitors spent an average of 72 minutes in forests in pre-pandemic times. Similar data in the UK Forest Research (2021) found most visitors (39–51%) stayed in forests up to one hour between 2014 and 2019. However, here again, it is unclear whether the pandemic induced the extension of forest visits in this study. Most respondents did not report an increased length of stay due to COVID-19. It is more likely that they practiced longer forest visits already before the pandemic.

When looking at motives, we observe slight changes with regard to pre-pandemic times. Similar to Ensinger et al. (2012) and UK Forest Research (2021), the motives "being in nature," "being outside," "physical exercise," and "good fresh air" are still important for visiting forests. Out of these, the motives of "physical exercise" and "being outside" gained in importance. The motives of "having a rest," "being away from everyday life," and "being able to switch off" kept the same degree of importance before and during the pandemic. These results hint at the fact that sporting facilities such as gyms were no longer available. Forests were one of the few remaining spaces for these activities and therefore were strongly sought-after. On the other hand, forests remained places to get away and relax, which they had been already before COVID-19, as reported by O'Brien et al. (2017). The overall structure of forest visitation motives, therefore, changed only slightly during COVID-19.

Another interesting finding relates to the subjective wellbeing of respondents. As reported above, the mean value of the sum score is at 13.99. This is below the value of a study that was done before COVID-19 by Brähler et al. (2007). Using a German representative sample, they found an average value for subjective wellbeing of 18.36. It might be assumed, therefore, that the pandemic could have negatively impacted perceived wellbeing. Ohlbrecht and Jellen (2021) and Ravens-Sieberer et al. (2022) report similar findings. Interestingly, however, this study shows that the subjective wellbeing of the respondents is higher than in a study by Baier and Kamenowski (2021), who reported an average value of subjective wellbeing of 11.30 for young people during the pandemic. Since the respondents of this study were frequent visitors to forests (over 80% visited forests once a week), this might hint at a positive effect of forests on wellbeing during COVID-19.

This assumption is further strengthened by looking at the regression analysis. Results show that longer stays in forests positively affected perceived wellbeing. Secondly, more frequent visits impacted on the extent to which wellbeing changed compared to pre-pandemic times. That is, respondents with frequent visits to forests underwent bigger changes than those who did not visit forests. If this finding is put into context with findings from Ohlbrecht and Jellen (2021), who stated that feelings of stress and exhaustion diminished in the first wave right after start of the pandemic, this might hint at the change being a positive one. This corresponds with Muro et al. (2022) who showed significant increases in positive affect, for example, vigor, friendship, and mindfulness in forest visitors practicing forest bathing during COVID-19.

However, it should be mentioned that the explanatory power of both regression models is low. For one, the regression coefficients are low. The length of stay in forests does explain a minor part of the changes in respondent subjective wellbeing only. Being more often in forests does explain changes in wellbeing to a higher degree, but it still is not a major explanatory variable. Furthermore, the value of R Square is low, indicating that a number of other factors, that were not part of this study, could have affected wellbeing during the pandemic. Furthermore, it needs to be noted that respondents, on average, had access to private green spaces, such as a garden. The results, therefore, are not fully generalizable. If the survey had been conducted among groups without access to green space, the importance of forest visits might have been even more evident. As the following section reveals, however, our findings are actionable in the realm of forest management.

Management Implications

COVID-19 has highlighted the importance of forests for human wellbeing. At the same time, the pandemic revealed the weakness of current forest management practices in the field of forest recreation. The increased numbers of forest visitors and their diverse activities posed challenges to forest administrations. To a previously unknown extent, forest spaces faced overcrowding and problematic behaviors like littering, vandalism, and illegal camping. Consequently, conflicts increased between forestry workers and visitors, but also between different recreational groups such as hikers and cyclists (Hansen et al., 2022). Moreover, forest management was confronted with new visitor profiles and a diversity of visitor expectations (Derks et al., 2020). Especially in nature protection areas, management reported being overwhelmed and hardly able to maintain their visitor services (Ferguson et al., 2022). COVID-19 and the related increase in forest visitors made clear that recreational issues need to be more comprehensively addressed in current forest management practices.

Sustainable forest management (SMF) aims at integrating environmental, economic, and social issues. Management of recreation, therefore, should be an integral part of forest management, alongside other forest uses like logging or hunting. In practice, however, recreational issues are of diminished importance. For example, in Europe the "Pan-European Criteria and Indicators for SMF" specify the implementation of sustainable forest management. The Criteria consists of 35 quantitative and 17 qualitative indicators. Interestingly, only a few relate to recreation, which are dispersed over several sections of the list. Against this, Pröbstl (2010) states: "Overall it appears that recreation, which was an explicitly defined component in the early implementations of multifunctional forestry, has now lost its special status" (p. 295). The same, to some extent, is true of forestry education. In many countries, recreational issues represent only a small part of forestry curricula. As far back as 2010, it has been argued that in undergraduate education in forest management, social sciences and humanities are not well represented (Vonhof, 2010). Another point is an economic one. Forests, in many cases, are managed from an economic point of view. That is, logging and wood harvesting are paramount, since they generate financial income, whereas recreation, by and large, does not. This is the background for the minimum attention paid to forestry in the field of recreation management, which was revealed by the pandemic. COVID-19, therefore, provides for a multitude of insights on forest management. In the following section, we discuss a few practical implications.

Taking into account that outcomes of forest visits like recreation are co-produced between forests as environmental spaces and visitors, at least three requirements for forest management arise. First is the requirement for a better understanding of which forest structures serve recreational aims best. Second is understanding human values and the practices forest visits are embedded in. Finally, more visitors in forests potentially means more conflicts, which calls for proactive plans of action. All three requirements call for a change in management perspective.

CES, as relational constructs that emerge from the interaction of environment and human beings, imply that different environmental structures might induce different CES. Some experts argue that green spaces with a high degree of structural and habitat diversity, but also with a high degree of accessibility and availability of water hold a particularly high level of CES potential (Bryce et al., 2016; Hegetschweiler et al., 2017; Paracchini et al., 2014). Forest management, and especially sylviculture in forests close to cities, needs to take this into account. This would mean creating forest structures according to recreational requirements instead of logging requirements. Forest structures with high CES potential could also be dedicated to recreational usages more frequently. All in all, forests, and especially city forests, will have to be managed more strategically as recreational spaces. This means taking into account CES and their co-production process more explicitly.

Practically, this means that place-specific data is required to understand exactly where recreational CES emerge (Hegetschweiler et al., 2017). One way to collect this data is to use a public participatory GIS, or geographic information system, which is a software program that helps analyze geographically referenced information (Dickinson & Hobbs, 2017). The advantage of this method is that CES can be recorded spatially. Examples of such instruments are OpenDataKit (ODK) or the Collector for ArcGIS. However, there are many other similar tools. Such instruments allow participants (who may be citizens, but could also include other stakeholders) to record parts of or points in landscapes, which are important for their recreation, in a geographically specific manner. This can be done with conventional devices such as a smartphone. For example, participants take georeferenced pictures with their smartphones, comment on them, and upload them to a database. In addition, deliberative methods, such as focus groups or citizen juries, may be used to capture different facets of human-nature interaction and to describe more precisely why and how visitors value certain forest structures (Dickinson & Hobbs, 2017).

We also note some disparity with regard to the population groups visiting forests. To change this and to secure access to forests for everyone, new forms of forest governance are needed. These should be more inclusive and participatory. Good examples come from urban green space planning. There, a communicative turn has taken place that shifts the roles and responsibilities of administration, politics, and citizens (Buijs et al., 2016; Fox-Kämper et al., 2018). Urban green spaces are now understood as urban green commons (Colding et al., 2013). These are physical green spaces in urban settings of diverse ownership that depend on collective organization and management and to which individuals and interest groups participating in management hold a rich set of rights, including rights to craft their own institutions and to decide whom they want to include in the management schemes (Colding et al., 2013).

The focus, therefore, is on active citizens who are able to organize themselves, mobilize resources, speak up in public, and represent their commons (Buijs et al., 2016). These active citizens define common property regimes which are social arrangements that regulate the maintenance and consumption of public green spaces (Colding et al., 2013).

To accomplish this, it is essential to involve a diversity of groups and to include their view-points and preferences. More precisely, this means (a) identifying social inequalities in forest visitation and taking them into account in planning and management, (b) incorporating different social perspectives, needs, and concerns into planning and management, (c) keeping an eye on existing and future trade-offs between forest planning and social goals (e.g., rise in prices for flats close to forests), and (d) planning forests in such a way that different groups gain access and use (Haase et al., 2017). The inclusion of diverse points of view has positive effects on the integrative power of forests as well: "Interviewees stressed that they identified more with their direct surrounding, creating a feeling of belonging and home if they can actively modify the environment or public green spaces" (Riechers et al., 2016).

Finally, to live up to the importance of green spaces for nature conservation, and to minimize conflicts, forest managers need to adopt an advanced plan of action. This means a need for precise information about visitor hotspots in forests. Forest management should know which forest areas are used by significant streams of visitors, when, and how. This data could be integrated with digital forest data in order to plan for forest management measures (e.g., logging). In this way, conflicts of use could be reduced. Since the number of inexperienced forest visitors increased during COVID-19, forest management should not assume basic rules of behavior are evident. Instead, group-specific communication with different messages is needed. New, modern forms of communication with forest visitors are necessary. Here, cooperation between forest management and tourism organizations is of great importance.

Conclusion

During the COVID-19 pandemic, forests played an important role in human wellbeing. This study shows that people visited forests more frequently during the first wave of COVID-19. Some of them extended their visits to forests and stayed there longer. These changes in forest visitation patterns contributed to wellbeing in a positive way. However, the regression coefficients are small, indicating that visiting forests longer and more often influenced wellbeing to a small degree only.

Nevertheless, when taking into consideration wellbeing as a cultural ecosystem service (CES) to which forests can make important contributions, forest administrations need to take recreational issues into account in a more comprehensive way. From a practical perspective, there are the requirements of a precise understanding of forest structures that suit recreation, the inclusion of a diverse range of societal groups into forest planning and management, as well as advanced formulation of strategies to prevent conflicts due to diverse user groups and expectations in forests.

The basic message, therefore, is that forests need to be managed more comprehensively with regard to recreational CES. This calls for a broader integration of recreation into daily routines of forest management, as well as into forest education curricula. Furthermore, reliable knowledge on existing and potential CES is needed. Here, a solid, georeferenced database could help, but also more inclusive and participatory forest governance. This would include strategic cooperation between forest management and adjacent societal stakeholders, such as individuals working in tourism or public welfare.

References

Baier, D., & Kamenowski, M. (2021). Jugend in Zeiten der Covid19- Pandemie Ergebnisse einer Wiederholungsbefragung von Jugendlichen im Kanton Zürich [Youth in times of Covid19. Results of a repeat survey of young people in the canton of Zurich]. *Fachhochschule Zürich*. Retrieved April 6, 2024 from https://digitalcollection.zhaw.ch/bitstream/11475/22094/3/2021_Baier-Kamenowski_Jugend-in-Zeiten-der-Covid19-Pandemie.pdf

Bamwesigye, D., Fialová, J., Kupec, P., Łukaszkiewicz, J., & Fortuna-Antoszkiewicz, B. (2021). Forest recreational services in the face of COVID-19 pandemic stress. *Land, 10*(12), 1347. https://doi.org/10.3390/land10121347

Beland Lindahl, K., & Westholm, E. (2011). Food, paper, wood, or energy? Global trends and future Swedish forest use. *Forests, 2*(1), 51–65. https://doi.org/10.3390/f2010051

Boyd, F., White, M., Bell, S., & Burt, J. (2018). Who doesn't visit natural environments for recreation and why: A population representative analysis of spatial, individual and temporal factors among adults in England. *Landscape and Urban Planning, 175*, 102–113. https://doi.org/10.1016/j.landurbplan.2018.03.016

Brähler, E., Mühlan, H., Albani, C., & Schmidt, S. (2007). Teststatistische Prüfung und Normierung der deutschen Versionen des EUROHIS-QOL Lebensqualität-Index und des WHO-5 Wohlbefindens-Index [Test statistical evaluation and standardization of the German versions of the EUROHIS-QOL quality of life index and the WHO-5 well-being index]. *Diagnostica, 53*(2), 83–96.

Brown, D. K., Barton, J. L., & Gladwell, V. F. (2013). Viewing nature scenes positively affects recovery of autonomic function following acute-mental stress. *Environmental Science & Technology, 47*(11), 5562–5569. https://doi.org/10.1021/es305019p

Bryce, R., Irvine, K., Church, A., Fish, R., Ranger, S., & Kenter, J. (2016). Subjective well-being indicators for large-scale assessment of cultural ecosystem services. *Ecosystem Services, 21*(B), 258–269. https://doi.org/10.1016/j.ecoser.2016.07.015

Buecker, S., Horstmann, K. T., Krasko, J., Kritzler, S., Terwiel, S., Kaiser, T., & Luhmann, M. (2020). Changes in daily loneliness for German residents during the first four weeks of the COVID-19 pandemic. *Social Science & Medicine, 265*, 113541. https://doi.org/10.1016/j.socscimed.2020.113541

Buijs, A. E., Mattijssen, T. J., Van der Jagt, A. P., Ambrose-Oji, B., Andersson, E., Elands, B. H., & Møller, S. M. (2016). Active citizenship for urban green infrastructure: Fostering the diversity and dynamics of citizen contributions through mosaic governance. *Current Opinion in Environmental Sustainability, 22*, 1–6. https://doi.org/10.1016/j.cosust.2017.01.002

Colding, J., Barthel, S., Bendt, P., Snep, R., van der Knaap, W., & Ernstson, H. (2013). Urban green commons: Insights on urban common property systems. *Global Environmental Change, 23*(5), 1039–1051. https://doi.org/10.1016/j.gloenvcha.2013.05.006

Derks, J., Giessen, L., & Winkel, G. (2020). Covid-19-induced visitor boom reveals the importance of forests as critical infrastructure. *Forest Policy and Economics, 118*, 102253. https://doi.org/10.1016/j.forpol.2020.102253

Dickinson, D. C., & Hobbs, R. J. (2017). Cultural ecosystem services: Characteristics, challenges and lessons for urban green space research. *Ecosystem Services, 25*, 179–194. https://doi.org/10.1016/j.ecoser.2017.04.014

DW (Deutsche Welle). (2021). *Most Germans feel pandemic has not drastically changed their lives, survey finds*. Retrieved March 8, 2022, from https://www.dw.com/en/most-germans-feel-pandemic-has-not-drastically-changed-their-lives-survey-finds/a-56427024

EEA. (2016). *European forest ecosystems—State and trends*. EEA Report No5. Retrieved March 8, 2022, from https://www.eea.europa.eu/publications/european-forest-ecosystems

Ensinger, K., Wurster, M., Selter, A., Jenne, M., Bethmann, S., & Botsch, K. (2012). Eintauchen in eine andere Welt—Untersuchung über Erholungskonzepte und Erholungsprozesse im Wald [Immersion into another world—Recovery concepts and recovery processes in forests]. *AFZ der Wald, 184*(3/4), 70–83.

Entringer, M., Kroeger, H., Schupp, J., Kühne, S., Liebig, S., Goebel, J., Grabka, M., Graeber, D., Kroh, M., Schröder, C., & Seebauer, J. (2020). *Psychische Krise durch Covid-19? Sorgen sinken, Einsamkeit steigt, Lebenszufriedenheit bleibt stabil [Mental crisis caused by Covid-19? Worries decrease, loneliness increases, life satisfaction remains steady]*. SOEP papers on Multidisciplinary Panel Data Research; 1087, Deutsches Institut für Wirtschaftsforschung (DIW), Berlin.

Ferguson, M. D., Lynch, M. L., Evensen, D., Ferguson, L. A., Barcelona, R., Giles, G., & Leberman, M. (2022). The nature of the pandemic: Exploring the negative impacts of the COVID-19 pandemic upon recreation visitor behaviors and experiences in parks and protected areas. *Journal of Outdoor Recreation and Tourism, 41,* 100498. https://doi.org/10.1016/j.jort.2022.100498

Fischer, A., & Eastwood, A. (2016). Coproduction of ecosystem services as human-nature interactions—An analytical framework. *Land Use Policy, 52*, 41–50. https://doi.org/10.1016/j.landusepol.2015.12.004

Fish, R., Church, A., Willis, C., Winter, M., Tratalos, J. A., Haines-Young, R., & Potschin, M. (2016a). Making space for cultural ecosystem services: Insights from a study of the UK nature improvement initiative. *Ecosystem Services, 21*, 329–343. https://doi.org/10.1016/j.ecoser.2016.09.017

Fish, R., Church, A., & Winter, M. (2016b). Conceptualising cultural ecosystem services: A novel framework for research and critical engagement. *Ecosystem Services, 21*, 208–217. https://doi.org/10.1016/j.ecoser.2016.09.002

Fox-Kämper, R., Wesener, A., Münderlein, D., Sondermann, M., McWilliam, W., & Kirk, N. (2018). Urban community gardens: An evaluation of governance approaches and related enablers and barriers at different development stages. *Landscape and Urban Planning, 170*, 59–68. https://doi.org/10.1016/j.landurbplan.2017.06.023

Grima, N., Corcoran, W., Hill-James, C., Langton, B., Sommer, H., & Fisher, B. (2020). The importance of urban natural areas and urban ecosystem services during the COVID-19 pandemic. *PLoS One, 15*(12), e0243344. https://doi.org/10.1371/journal.pone.0243344

Haase, D., Kabisch, S., Haase, A., Andersson, E., Banzhaf, E., Baró, F., Brenck, M., Fischer, L. K., Frantzeskaki, N., Kabisch, N., Krellenberg, K., Kremer, P., Kronenberg, J., Larondelle, N., Mathey, J., Pauleit, S., Ring, I., Rink, D., Schwarz, N., & Wolff, M. (2017). Greening cities—To be socially inclusive? About the alleged paradox of society and ecology in cities. *Habitat International, 64*, 41–48. https://doi.org/10.1016/j.habitatint.2017.04.005

Hansen, A. S., Beery, T., Fredman, P., & Wolf-Watz, D. (2022). Outdoor recreation in Sweden during and after the Covid-19 pandemic—Management and policy implications. *Journal of Environmental Planning and Management*. 66, 1472–1493. https://doi.org/10.1080/09640568.2022.2029736

Hartig, T., van den Berg, A. E., Hagerhall, C. M., Tomalak, M., Bauer, N., Hansmann, R., Ojala, A., Syngollitou, E., Carrus, G., van Herzele, A., Bell, S., Podesta, M. T. C., & Waaseth, G. (2011). Health benefits of nature experience: Psychological, social and cultural processes. In Nilsson, K., et al. (Eds). *Forests, Trees and Human Health*. (p. 127–168). Springer: Dordrecht. https://doi.org/10.1007/978-90-481-9806-1_5

Hegetschweiler, K. T., de Vries, S., Arnberger, A., Bell, S., Brennan, M., Siter, N., Olafsson, A. S., Voigt, A., & Hunziker, M. (2017). Linking demand and supply factors in identifying cultural ecosystem services of urban green infrastructures: A review of European studies. *Urban Forestry & Urban Greening, 21*, 48–59. https://doi.org/10.1016/j.ufug.2016.11.002

Hunziker, M., Frick, J., von Lindern, E., & Bauer, N. (2013). Das Verhältnis der Bevölkerung zum Wald [The population's relation to the forests]. *Bündnerwald, 66*(1), 30–34.

Jay, M., & Schraml, U. (2009). Understanding the role of urban forests for migrants—Uses, perception and integrative potential. *Urban Forestry & Urban Greening, 8*(4), 283–294. https://doi.org/10.1016/j.ufug.2009.07.003
Korpela, K., Borodulin, K., Neuvonen, M., Paronen, O., & Tyrväinen, L. (2014). Analyzing the mediators between nature-based outdoor recreation and emotional well-being. *Journal of Environmental Psychology, 37*, 1–7. https://doi.org/10.1016/j.jenvp.2013.11.003
Lee, J., Park, B. J., Tsunetsugu, Y., Ohira, T., Kagawa, T., & Miyazaki, Y. (2011). Effect of forest bathing on physiological and psychological responses in young Japanese male subjects. *Public Health, 125*(2), 93–100. https://doi.org/10.1016/j.puhe.2010.09.005
Lupp, G., Förster, B., Kantelberg, V., Markmann, T., Naumann, J., Honert, C., Koch, M., & Pauleit, S. (2016). Assessing the recreation value of urban woodland using the ecosystem service approach in two forests in the Munich Metropolitan region. *Sustainability, 8*(11), 1156. https://doi.org/10.3390/su8111156
MA (Millennium Ecosystem Assessment). (2005). *Ecosystems and human well-being: Synthesis*. Retrieved March 8, 2022, from https://www.millenniumassessment.org/documents/document.356.aspx.pdf
Muro, A., Feliu-Soler, A., Canals, J., Parrado, E., & Sanz, A. (2022). Psychological benefits of Forest Bathing during the COVID-19 pandemic: A pilot study in a Mediterranean forest close to urban areas. *Journal of Forest Research, 27*(1), 71–75. https://doi.org/10.1080/13416979.2021.1996516
O'Brien, L., De Vreese, R., Kern, M., Sievänen, T., Stojanova, B., & Atmiş, E. (2017). Cultural ecosystem benefits of urban and peri-urban green infrastructure across different European countries. *Urban Forestry & Urban Greening, 24*, 236–248. https://doi.org/10.1016/j.ufug.2017.03.002
O'Brien, L., & Morris, J. (2014). Well-being for all? The social distribution of benefits gained from woodlands and forests in Britain. *Local Environment, 19*(4), 356–383. https://doi.org/10.1080/13549839.2013.790354
Ohlbrecht, H., & Jellen, J. (2021). Unequal tensions: The effects of the coronavirus pandemic in light of subjective health and social inequality dimensions in Germany. *European Societies, 23*, S905–S922. https://doi.org/10.1080/14616696.2020.1852440
OIA (Outdoor Industry Association). (2021). *Outdoor participation trends report*. Outdoor Industry Association. Retrieved March 8, 2022, from https://outdoorindustry.org/resource/2021-outdoor-participation-trends-report/

Paracchini, M. L., Zulian, G., Kopperoinen, L., Maes, J., Schägner, J. P., Termansen, M., Zandersen, M., Perez-Soba, M., Scholefield, P. A., & Bidoglio, G. (2014). Mapping cultural ecosystem services: A framework to assess the potential for outdoor recreation across the EU. *Ecological Indicators, 45*, 371–385. https://doi.org/10.1016/j.ecolind.2014.04.018

Pleasant, M. M., Gray, S. A., Lepczyk, C., Fernandes, A., Hunter, N., & Ford, D. (2014). Managing cultural ecosystem services. *Ecosystem Services, 8*, 141–147. https://doi.org/10.1016/j.ecoser.2014.03.006

Pröbstl, U. (2010). Lessons learned, trends and strategies for the future. In: Pröbstl, U., Wirth, V., Elands, B., Bell, S. (Eds). *Management of Recreation and Nature Based Tourism in European Forests*. (p. 287–298). Springer: Berlin, Heidelberg. https://doi.org/10.1007/978-3-642-03145-8_9

Ravens-Sieberer, U., Kaman, A., Erhart, M., Devine, J., Schlack, R., & Otto, C. (2022). Impact of the COVID-19 pandemic on quality of life and mental health in children and adolescents in Germany. *European Child & Adolescent Psychiatry, 31*(6), 879–889. https://doi.org/10.1007/s00787-021-01726-5

Rice, W. L., Mateer, T. J., Reigner, N., Newman, P., Lawhon, B., & Taff, B. D. (2020). Changes in recreational behaviors of outdoor enthusiasts during the COVID-19 pandemic: Analysis across urban and rural communities. *Journal of Urban Ecology, 6*(1). Article juaa020. https://doi.org/10.1093/jue/juaa020

Riechers, M., Barkmann, J., & Tscharntke, T. (2016). Perceptions of cultural ecosystem services from urban green. *Ecosystem Services, 17*, 33–39. https://doi.org/10.1016/j.ecoser.2015.11.007

Stålhammar, S., & Pedersen, E. (2017). Recreational cultural ecosystem services: How do people describe the value? *Ecosystem Services, 26*, 1–9. https://doi.org/10.1016/j.ecoser.2017.05.010

UK Forest Research. (2021). *Forestry Statistics 2021. Compendium of statistics about woodland, forestry and primary wood processing in the United Kingdom.* Retrieved March 8, 2022, from https://www.forestresearch.gov.uk/tools-and-resources/statistics/forestry-statistics/forestry-statistics-2021

Vonhof, S. (2010). Deficiencies of undergraduate forestry curricula in their social sciences and humanities requirements. *Journal of Forestry, 108*, 413–418. https://doi.org/10.1093/jof/108.8.413

Weinbrenner, H., Breithut, J., Hebermehl, W., Kaufmann, A., Klinger, T., Palm, T., & Wirth, K. (2021). "The Forest Has Become Our New Living Room"—The critical importance of urban forests during the COVID-19 pandemic. *Frontiers in Forests and Global Change, 4*. https://doi.org/10.3389/ffgc.2021.672909

White, E., Bowker, J. M., Askew, A. E., Langner, L. L., Arnold, J. R., & English, D. B. K. (2016). *Federal outdoor recreation trends: Effects on economic opportunities*. Retrieved April 6, 2024 from https://www.fs.usda.gov/research/treesearch/53247#

WHO. (1998). *Wellbeing measures in primary health care. WHO Regional Office for Europe*. Retrieved April 6, 2024 from https://iris.who.int/bitstream/handle/10665/349766/WHO-EURO-1998-4234-43993-62027-eng.pdf?sequence=1&isAllowed=y

6

Staycations: A Sustainable Tourism Solution?

Gianna Moscardo

Introduction

> *People report feeling awe in response to more mundane things: when seeing the leaves of a Gingko tree change from green to yellow, in beholding the night sky... Intriguingly, each burst of daily awe predicted greater well-being* (Keltner, 2016)

In this quote, Keltner, a professor of psychology, describes the potential of finding value through a deeper engagement with the immediate world around us, suggesting the possibility that distance, difference, and departure from our everyday lives may not be as important to tourism as we have often assumed. In the years prior to the COVID-19 pandemic, much of the discussion about sustainable tourism focused on the need for degrowth in tourism (Hall, 2011). Hall (2011) asked whether it was necessary for tourists to travel so far and so often in order to gain the personal benefits they associated with holidays away from home. Degrowth

G. Moscardo (✉)
James Cook University, Townsville, QLD, Australia
e-mail: gianna.moscardo@jcu.edu.au

T. Walker et al. (eds.), *Sustainable Tourism*,
https://doi.org/10.1007/978-3-031-43528-7_6

was about limiting tourist consumption and impact to more manageable levels and included three types of change in tourist practices: less tourism overall by encouraging everyone to travel less often, slower tourism combining less travel (especially by plane) with more intensive engagement in a single destination region, and more staycations or proximity tourism (Molz, 2009; Ram et al., 2013; Salmela et al., 2021; Vogt & Andereck, 2018).

Demand for staycations increased prior to the COVID-19 pandemic, albeit slowly and from a small base of tourists seeking to slow down, retreat, and focus on simpler, less conspicuous, and more sustainable travel options (Sørensen & Hjalager, 2020). Emerging discussions of more sustainable destination marketing also highlighted the importance of local and intraregional tourists as a market segment that was likely to have less of a carbon footprint from significantly reduced travel distance and that was likely to spend more locally (Kastenholz et al., 2018; Moscardo & Murphy, 2016). Arguably, locals spending their money within the local economy is better than taking it out of the region. This latter theme was one used extensively by destinations when travel restrictions due to COVID-19 began to ease, with people being encouraged to take a holiday locally to provide support to struggling tourism businesses. COVID-19 generated a significant increase in attention to staycations both because for many people, they were the only option available for long periods of time and because they have often been portrayed as safer than longer distance travel.

In these discussions of degrowth, staycations have been seen as contributing to tourism sustainability in two main ways:

- By limiting long-distance travel and thus reducing carbon emissions, pressure on transport infrastructure, the need for expanding global transport systems, and the consequential impacts of these expansions on the physical environment, and
- By keeping expenditure within the local destination region, thus limiting some of the negative consequences of economic leakage.

These two key elements give us some insights into how a staycation might be defined. Draper (2017, p. 1178) provides a rare formal academic definition of a staycation as "a shortened trip closer to the residence of the participants." Media and tourism promotional discussions of the concept also focus on being a tourist, that is using commercial tourism products and services including tours and activities typically focused on extra-regional visitors. Therefore, this chapter defines a staycation as the phenomenon of taking a holiday close to home which might be accessed with public transport that specifically includes engaging in commercial tourism activities and services. Staycations typically involve residents traveling within a tourist destination region and as such are usually part of domestic tourism. There may be some exceptions to this, for example, where people live close to an international border, it may be possible to have an international staycation that is very close to home. It must also be recognized that perceptions of what constitutes a shorter trip are likely to differ across locations. In some countries with smaller, more dispersed populations, a staycation may involve more travel than for those who live in more closely and densely populated areas. The critical components of a staycation require limiting the physical distance traveled, staying within the destination region, and using local commercial tourism options. Staycations may also be seen as overlapping with some recreational activities such as attending sports events and engaging in outdoor recreational activities such as fishing, hunting, bird watching, and hiking. The differences here are of degree of interaction with commercial tourism businesses, with recreationists more likely to be relatively independent, and the focus of the activity, with recreationists more likely to focus greater attention on the target recreational pursuit.

Staycations emerged as a named form of tourism at the end of the first decade of this century in response to the global financial crisis (Besson, 2017). The forces encouraging this new form of tourism focused on cost savings, ease of planning, and keeping money within local communities to help them respond to the recession (Fox, 2009; Molz, 2009). For many places, COVID-19 made staycations the only option for tourists for long periods of time and so the last few years have offered an opportunity to explore the benefits and understand the processes connected to this sustainable tourism solution. After briefly reviewing current tourism provider representations of staycations and the available academic research on staycations, this chapter takes a systems thinking approach (Betley et al., 2021) to analyze staycations. Discussions of sustainability and other complex issues facing contemporary societies have consistently stressed the need to find new approaches to both problem solving and research and knowledge generation with systems thinking emerging as one of these new approaches (Lich et al., 2017). Based on a review of system thinking definitions, Arnold and Wade (2015) offer a now widely used definition. In this definition, systems thinking is defined as a set of "analytic skills used to improve the capability of identifying and understanding systems, predicting their behaviors, and devising modifications to them in order to produce desired effect" (Arnold & Wade, 2015, p. 675). The goal of research in systems thinking is not to describe in detail all the features of a static system, but instead to understand the system as a working dynamic entity in sufficient depth to identify the core interactions or leverage points where change can be most easily made (McCool et al., 2015). Using this approach, this chapter maps out the core elements of staycations and identifies the key leverage points in the tourism system that support a shift toward staycations (McCool & Kline, 2020). Then, this chapter uses these key leverage points to suggest strategies that could be used to encourage greater use of staycations for more sustainable tourism.

Current Tourism Provider Representations of Staycations

An examination of travel industry reports and commentary in travel industry magazines and websites from 2020 and 2021 revealed five key themes in discussions of staycations both specifically and as part of a move to domestic tourism more generally. Firstly, the United Nations World Tourism Organization's (UNWTO) (2020) review of national responses to COVID-19 restrictions on international travel noted that two main approaches were taken to encourage domestic tourism—domestic promotional campaigns and financial incentives. In those countries that eased domestic travel restrictions earlier in the pandemic national and regional tourism agencies offered various levels of financial incentives for residents to engage in domestic travel including discounted airfares, travel vouchers, and direct payments to tourism providers. Examples included Indonesia, Malaysia, Singapore, Thailand, Australia, at both the state and national level, New Zealand, Hong Kong, Norway, Sweden, Spain, and Egypt. These travel discounts and financial incentives were often supported by promotional campaigns with slogans like *Holiday Here This Year, Love Great Britain*, in the United Kingdom, and *Less Travel Time, More Family Time*, in the Australian state of Queensland. Secondly, in countries with internal travel restrictions, these campaigns encouraged staycations by highlighting their advantages including lower costs, less time spent on travel and more on holiday activities, being family friendly, supporting the local economy, and being an opportunity to better understand your home place (Edgar, 2020; PeekPro, 2020; Sudevan, 2020). Thirdly, these campaigns often focused on how people could find experiences closer to home that matched those typically sought overseas (Escape, 2021a) and that there was much benefit to be found from looking at one's own region and seeking the exotic beneath the mundane (Escape, 2021b). This was consistent with industry discussions of how tourism providers had to rethink the design of their experiences, recognizing that what was interesting for international visitors was not necessarily going to capture the attention of *staycationers* (Parsons, 2021). Fourthly, there was recognition that for countries where more residents traveled overseas than international visitors arrived pre-COVID-19, such as Australia, Canada, and Germany, this shift to domestic tourists and staycations could create greater returns from tourism than usual. In countries with the opposite balance of outbound to inbound tourism, such as Spain, Japan, and Iceland, staycations were unlikely to maintain the sector at pre-COVID-19 levels (Ali, 2020; KPMG, 2021). This latter problem was especially true for less affluent international tourism-dependent destinations, such as Barbados, Jamaica, the Seychelles and Maldives, and Fiji, where there is a smaller number of residents who have limited capacity to take even local holidays. This type of reliance on international tourism was linked to the final theme of waiting to return to normal (Dinev, 2020). Some countries used campaigns built around the concept of encouraging potential tourists to dream of a holiday to the destination and save for when that holiday would be possible. Switzerland's *Dream*

Now, Travel Later, and *Visit Maldives Later* campaigns are examples of this (Bolchinova, 2020). As this chapter is written at the beginning of 2022, when international travel is no longer restricted in most places around the world, many nations have shifted their focus away from staycations and domestic travel toward encouraging the return of international tourists.

Academic Research into Staycations

Although there have been calls for increased staycations as a sustainable solution for tourism and, in the last two years, there has been considerable media and travel industry discussion of staycations as a solution to COVID-19-induced challenges for both tourist destinations and cautious tourists, there has been surprisingly little academic research into staycations (Salmela et al., 2021). The available academic research on staycations can be classified into two main areas—healthcare research into the benefits of taking breaks from everyday work and life and tourism research into staycations.

It is often assumed in tourism-related academic discussions that traveling on holidays generates many health-related benefits for tourists, and where this assumption is supported by evidence (cf., Chen & Petrick, 2016), it is often limited and biased toward studies that report some health benefits, ignoring those that report no benefits. Other tourism researchers, even when citing studies that do report benefits, often omit the detail in these studies, which typically report very small and fleeting health benefits (de Bloom et al., 2009; Reizer & Mey-Raz, 2019; Sonnentag et al., 2022). Tourism researchers in general have also largely ignored the considerable body of medical literature highlighting the negative health impacts of travel on tourists (Herman & Patel, 2018; Korzeniewski, 2020; Long & Flaherty, 2018). Other tourism papers report benefits but can be seen as suffering from issues of assumed causality and survivor bias (cf., Mélon et al., 2018). In the former, case studies that measure overall health and/or wellbeing and then look at correlations with extent of travel often falsely assume that more travel causes better health, despite the very real possibility that better health allows for more travel. The second case, survivor bias is a consistent issue across many tourism studies where the positive responses of those who chose to travel or continue to travel are seen as the result of travel experiences and it is assumed, without any comparative evidence, that those who do not travel will not report these benefits (Kessler et al., 2015; Moscardo, 2019).

Despite the limitations reported in the previous paragraph, there is evidence that taking a break from everyday life can benefit mental and physical wellbeing. Very few studies in the areas of work recovery, stress management, health, and organizational studies compare a holiday at home to a holiday taken away from home (de Bloom et al., 2017), and even fewer compare either of these two options to a staycation. In general, where research compares leisure time spent at home versus that used for a holiday taken at a location further away from home, the research reports either no difference in benefits (cf. Blank et al., 2018) or small differences in benefits in favor of being away from home (cf. de Bloom et al., 2017), although it must be stressed this is based on very limited evidence. Research into the benefits of time taken away from work and everyday stressors does, however, provide us with insights into the factors that contribute to improved wellbeing outcomes, and this highlights variables likely to contribute to staycation benefits. Five interrelated concepts have been identified as important for recovery benefits:

- Deliberate detachment from everyday work routines and commitments including domestic chores (de Bloom et al., 2017; Smyth et al., 2020; Sonnentag et al., 2022),
- Finding and engaging in specific relaxation activities such as sleeping, reading, gaming, spa treatments, and physical activity such as sport and exercise (Sonnentag et al., 2022),
- Mastery, which refers to time spent practicing, learning, and improving various creative, sporting, or other skills (Sonnentag et al., 2022),
- Control, which refers to being able to exercise self-determination and agency and minimize opportunity costs (Cardini & Freund, 2021; Kuykendall et al., 2021; Sonnentag et al., 2022), and
- Opportunities to connect with family and friends (Kuykendall et al., 2021).

Staycations offer all these characteristics and as they are taken in familiar environments, it could also be argued that they are better than holidays taken further away, offering more time for these recovery opportunities and greater control. The major challenge for a staycation is detachment, especially from domestic chores and work communications in the COVID-19 era where work and home boundaries have become blurred (White, 2021). There is evidence, however, that the use of explicit detachment strategies (Sonnentag & Niessen, 2020) such as a smartphone app (Smyth et al., 2020) or simply acting as if they were a tourist on holiday elsewhere (West et al., 2021) can be effective.

This evidence from studies of the health and wellbeing impacts of vacations suggests that staycations are likely to be as, or more, effective for improving individual wellbeing than distanced holidays. This is because the benefits of a vacation lie mainly in the break in routine and freedom from everyday work and life pressures and constraints (de Bloom et al., 2009, 2017). These benefits are based on the break not the location visited. These studies do, however, focus on inner directed holiday forces such as individual health, mastery, and links to important others. This work does not consider outer directed forces for more distant holiday travel, such as social status and conspicuous consumption. Issues of affluence and consumption have not been addressed in this area but are likely to be both critical facilitators and important self-esteem benefits of more distant holiday options.

The perceived benefit of a staycation is the first of two core themes in tourism research which focuses on this type of vacation. Descriptive discussions of staycations report that tourists see them as more affordable, making it easier to have more frequent breaks, easier to plan and manage with less time wasted on travel to and from the destination, more family/relationship focused, and safer (Besson, 2017; Moya Calderón et al., 2021; Fox, 2009; James et al., 2017). More detailed analyses have focused on the pathways linking features of tourists to their activities and features of the staycation experience to various tourist wellbeing outcomes. A quantitative analysis by Lin et al. (2021), for example, found that tourists seeking escape, novelty, and nostalgia in a staycation, who engaged with more tourism elements in the local region, were more likely to report restorative and life satisfaction benefits from a staycation. A qualitative approach by Besson (2017) also found that more engagement with the local environment in terms of more actively seeking to find beauty, fun, and adventure closer to home was associated with greater eudemonic benefits. This work confirms the findings of health studies that, under the right conditions, it is possible to generate the same (or more) personal benefits from a holiday spent at or closer to home, as one based on more distant travel.

The second theme in tourism research focuses on how staycations challenge our concepts of what a holiday or vacation is, highlighting the importance of distance and (im)mobility in our assumptions about what a person must do to be a tourist. Molz's (2009) discussion of staycations as a different type of tourism mobility in which pace and distance are separated as different dimensions reports critics of staycations. These critics argue that to recover or escape from the pressures of a fast-paced modern life, a holiday needs to be taken at a distance from home and that holidays must be about movement, not stillness. In this conception, staycations as a response to the global financial crisis were seen as a stopgap for people "waiting for the world to return to normal" (Molz, 2009, p. 282). Molz (2009) also notes the inherent contradiction in the claim that the answer to dealing with pressures of a modern life characterized by speed is to travel long distances quickly. This contradiction and the sense that staycations may be acceptable while we wait for a change in the world are particularly relevant to contemporary post COVID-19 circumstances and are a cause for concern for those arguing for staycations as a long-term sustainable solution for tourism. Not everyone is, however, pessimistic, with Besson (2017) noting that, like Keltner in the opening quote of this chapter, there is value to be found in mundane things. Edelheim et al. (2020) also argue that changes in how we negotiate work and home boundaries imposed by the COVID-19 pandemic may help change how we view work, home, and leisure boundaries. Salmela et al. (2021, p. 54) argue that the meaning of tourism can "be actively interpreted, modified and renegotiated." This idea of changing fundamental understandings of a phenomenon as a pathway to sustainability is at the core of systems thinking applied to sustainability issues.

An Overview of Systems Thinking

A common element of systems thinking is to recognize that observable events, the focus of much research, are the top of a systems iceberg reflecting patterns/trends which emerge from the bulk of the iceberg below the waterline including system structures and drivers which, in turn, are based upon dominant social paradigms, representations, or worldviews (Kim, 1999). Many existing academic endeavors focus on mapping out, in detail, the top two layers of this model, hoping that this will reveal at least part of the system in the levels below (Inghels, 2020). Systems thinking argues for less detail at each level and more focus on how the levels are linked (Abson et al., 2017). The process is less concerned with describing the system than with changing it (Arnold & Wade, 2015; Knight et al., 2019; Moscardo, 2022).

Table 6.1 demonstrates how these four levels of analysis in systems thinking can be used to both summarize key conclusions from the available literature on staycations and to suggest how the levels are interlinked. While recent global travel conditions arising from COVID-19 made staycations a major element of tourism and there is evidence that they can offer not only sustainability benefits for tourism, but also personal benefits for tourists, both practice and research discussions highlight the widespread view that staycations are not seen as "real" tourism. There are suggestions that current tourism systems favor international travel and predominant worldviews see staycations as a temporary option until global travel resumes post-COVID-19 restrictions. The tourism sustainability challenge is to find the leverage points in current tourism systems that could be used to make staycations a preferred travel option.

A Systems Model of Staycations

Table 6.1 suggests that while staycations may offer a viable sustainability solution in tourism, the current system is unlikely to encourage their uptake, especially as COVID-19 travel restrictions disappear. The challenge then becomes to identify leverage points in the system that could be

Table 6.1 Levels of analysis in a systems thinking approach to staycations

Levels of analysis	Application to staycations
Surface events	Global recession (2007) and global pandemic (2020) drive an increase in staycations as the only options available for many tourists
Patterns and trends	Focus by tourists on safety, low risk, family and social connections, and personal wellbeing and recovery from stress as focus of travel rather than social status, conspicuous consumption, and novelty
	Evidence that with assistance, staycations can meet personal health and wellbeing needs
	Arguments that staycations are a way to improve tourism sustainability through reducing environmental costs of travel and increasing economic benefits for destination regions
	Recognition of the challenges of providing appropriate experiences for local residents in destination systems usually focused on external tourists and destinations where residents may not even be able to afford local holidays
System structures and drivers	Tourism providers campaigning for staycations as a safe and rewarding, but temporary, option while travel restrictions remain
	Focus is on personal wellbeing benefits for tourists and economic benefits for local regions
	Destination structures and institutions remain driven by a focus on extra-regional visitors
Predominant worldviews	"Real" tourist experiences require distance to meet tourist needs and support the commercial travel sector
	Staycations are not "real" tourist experiences as tourism is fundamentally an international phenomenon

changed to shift tourists and tourism providers perspective of staycations as the most common and desirable form of tourist experience. Moscardo (2019) summarizes the literature on system models by identifying their main features including:

- Critical elements, major actors, and key institutions,
- Interconnections, pathways, and/or causal relationships that link these elements,
- External forces exerting pressures on the system,

- Emergent properties or outcomes from the functioning of the system, and
- Leverage points where intervention is most likely to achieve change.

Figure 6.1 is a simple systems model highlighting the critical elements and dominant forces in the current tourism system. Predominant worldviews or representations of tourism focus heavily on international and long-distance travel highlighting novelty, escape from home and everyday life, and conspicuous consumption (Hall, 2011; Moscardo, 2019; Ram et al., 2013). This is communicated to potential tourists who then look

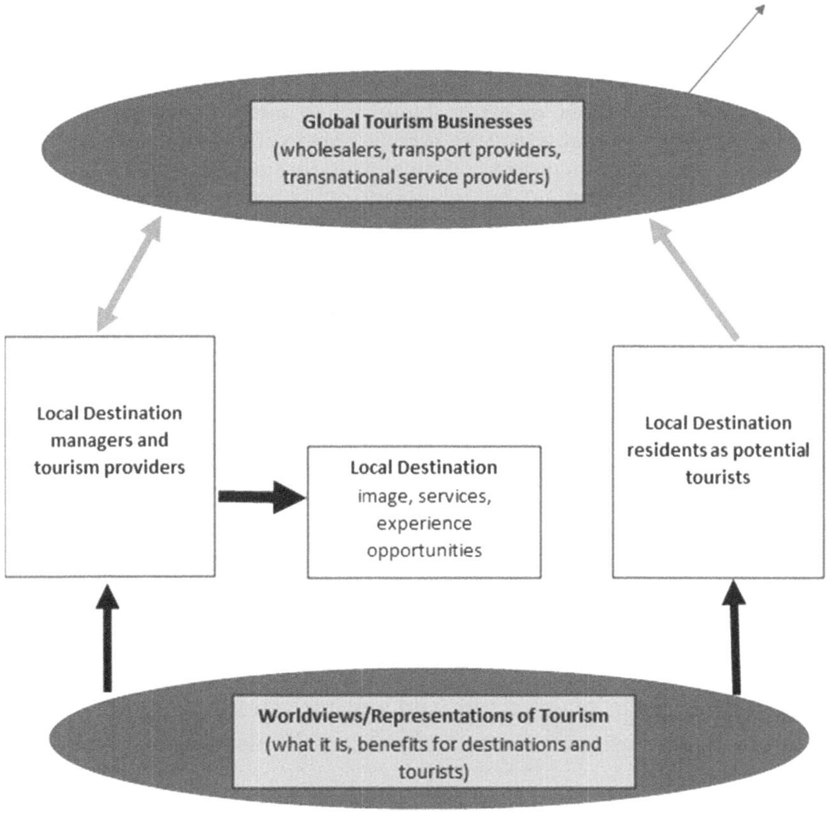

Fig. 6.1 A simple systems model of current tourism

to the global tourism businesses for holiday opportunities. These global tourism businesses are dependent on international travel, seeing destinations as interchangeable resources to support their revenue and with little vested interest in individual destination wellbeing or in changing the system (Brouder, 2020; Moscardo, 2019). Local residents are then directed away from their own region as a destination.

Many local destination managers and tourism providers mirror potential tourists in accepting the dominant worldview of tourism and looking toward global tourism businesses for their customers. This is evident in a common destination management organization (DMO) focus on international market segments, generating new and more direct transports links, especially for air traffic and attracting external investment in tourism infrastructure (Morgan, 2012; Moscardo, 2011). These destination managers and tourism providers then focus their promotion and experience offerings to suit these more distant markets. There are few direct links between DMOs and local destination residents. COVID-19 travel restrictions in many places cut the links between local destination managers and residents, and the global tourism business sector, thus disrupting these flows of tourists in and out of destination regions. The review of business and academic discussions about staycations in the last few years that most assumed, correctly, that this disruption would be temporary because it did not challenge the worldviews or representations of tourism, nor significantly alter the way destination tourism providers offer experiences, nor create more permanent links between destination managers, tourism providers, and residents. This suggests several leverage points to focus on for change.

Figure 6.2 highlights some of the changes needed in these leverage points for the overall system to support staycations in becoming a major form of tourism consumption. Firstly, change is needed in the way both potential tourists and destination tourism managers and providers think about what tourism is and what its benefits are for all parties involved. Secondly, local destination managers and tourism providers need to revisit the experience opportunities and services they provide to include locals and encourage them to become *staycationers*. Thirdly, stronger and more direct links between local destination tourism managers and providers and local residents need to be created.

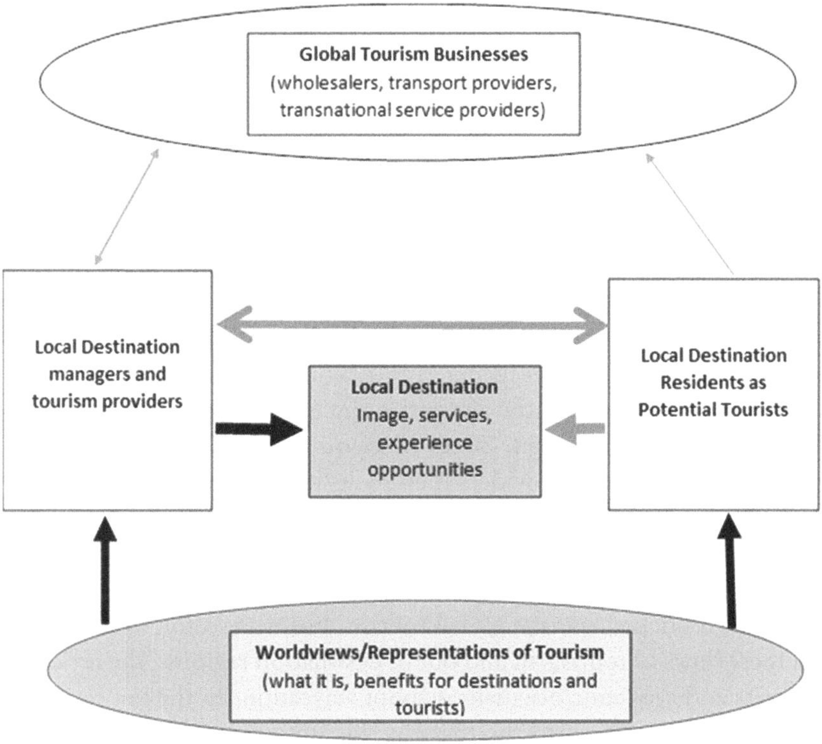

Fig. 6.2 A simple systems model of tourism supporting staycations

Strategies to Change Key Leverage Points

Stronger and more direct links between local destination tourism managers and providers and residents is the easiest of the three leverage points to address. An extension or revival of government financial incentives for taking a holiday or vacation closer to home is a simple but costly strategy for building and strengthening direct connections between residents and tourism providers within a destination. For many destinations, however, the likelihood of this being politically attractive is unlikely given the large amounts of funding that many governments have already invested in tourism businesses through COVID-19 response and recovery programs. This does not preclude individual tourism businesses from offering incentives for residents. A less immediate, but very important, option for building resident tourism business connections is changing the structures of tourism governance at a destination level to incorporate more options for residents to be actively engaged in tourism governance processes, thus increasing awareness of tourism as a local socio-cultural economic activity. Getting more resident empowerment in tourism may also alter the way destination managers think about their research, marketing, and experience design activities.

Conducting more research on resident tourism decision-making and holiday motivations, as well as expectations and evaluations of staycations taken during the COVID-19 pandemic could not only offer a mechanism for communication between the two groups, but also directly inform changes to the second leverage point of the experiences offered to shift from catering mostly to extra-regional tourists. While affordability, ease of travel, and less time spent on travel to the destination are obvious benefits of staycations, these are unlikely to be the factors that matter when people consider staycations against more distant travel options. If they were, staycations would already be a more popular option. Rather, marketing campaigns for staycations need to explicitly focus on the health, wellbeing, and social connection benefits that can be more easily obtained from a staycation and on the positive sustainability aspects especially contributing to local economies and reducing greenhouse gas emissions. Table 6.2 summarizes options for changes in tourist experiences linked to the demonstrated benefits of staycations and the factors that contribute to these benefits.

Many tourism operators focus on storytelling for extra-regional visitors, with a particular emphasis on telling the stories of the destination to visitors from further away, especially through visits to significant sites. As residents likely already know these local stories, tour operators will have to change their emphasis to telling different stories or lesser-known stories about these places than those normally presented to tourists and to providing opportunities for *staycationers* to create and live their own stories. Staycations mean that tourism becomes less about traditional storytelling and more about offering places and activities that tourists can use to write and star in their own personal stories (Moscardo, 2020). This is the essence of co-creation. While local tourism providers will need

Table 6.2 Directions for reorienting tourist experiences to include local visitors

Factor contributing to benefits	Changes to tourist experience opportunities
Deliberate detachment	Apps and programs provided by tourism providers to support detachment Building immersion and active engagement into experiences
Relaxation	Target experiences on relaxation techniques and/or structured outdoor nature activities
Mastery	Create learning and creativity activities focused on developing skills and encouraging individual expression
Control	Giving opportunities for active participation, decision-making, and personalization of experiences
Relationships and social connections	Building social and participatory elements into tourist experiences

training and support for this shift, it could also have benefits in terms of enhancing social capital amongst residents, creating more support for tourism amongst residents, and providing more rewarding and attractive experiences for extra-regional tourists who often seek opportunities to meet and interact with locals (Wei et al., 2019).

Changes in the way destinations think about and offer tourist experiences would begin the process of tackling the hardest leverage point to change: the worldviews of tourism. A major challenge here is that for many tourists, travel is a form of conspicuous consumption, even if not all tourists will admit to this, and that motivation is unlikely to be addressed by staycations. Like many dominant representations that are relevant to sustainability changes, shifting widely shared and socially embedded worldviews requires time and multiple public education campaigns across a variety of sectors. Sharing and supporting such campaigns by sustainability-oriented tourism organizations and businesses is important. It is also important to link staycations to other sustainability actions. For academics, it is important to include discussions about degrowth and issues linked to tourism consumption in education programs about tourism and sustainability.

Issues and Future Directions

Taking a systems thinking approach which emphasizes a different way to map out and analyze complex problems, such as sustainability, could lead us to identifying different ways to think about solutions to these problems. If we argue that widespread adoption of staycations is an important strategy to improve tourism sustainability, then identifying the important leverage points in the tourism system is a critical step to developing these strategies. Traditional thinking focusses on understanding the elements of the system in more detail, especially local residents and tourism providers, rather than interactions and underlying assumptions. Thus, more traditional approaches to tourism research to support staycations are likely to focus on exploring tourists' perceptions and evaluations of, and interest in, staycations as a specific form of tourism. Matching this in the world of tourism practitioners, such as destination planners and marketers, would be calls for new promotional campaigns targeted at local visitors.

While these might be useful to some extent, a comparison of the systems model for current tourism with a system model of tourism that supports staycations identifies three key leverage points that are quite different to those in the more traditional approach. These are:

- Changing the nature of the experiences offered and promoted at the destination, especially focusing on co-creation of new stories of the destination, rather than changing the advertising of the same experiences,
- Engaging and empowering destination residents in tourism governance by giving them more exposure to, personal investment in, and co-creation of tourism experience opportunities, and
- Challenging the underlying assumptions that tourism is about speedy travel to distant and exotic locations.

While this chapter argues for staycations as a tourism sustainability solution, it is important to note that staycations are not entirely free of sustainability challenges and issues and are unlikely to be a sole solution to sustainability. There is anecdotal evidence that COVID-19-induced staycations have caused negative impacts resulting from sudden and intense pressures on natural environments. If staycations are linked to second homes in some destinations, then the issues associated with second home ownership will also have to be considered. Staycations are also limited to those who can afford to pay for holidays, and while cheaper than other tourist options, they are still only available to those able to afford them. Staycations are currently not a viable replacement for those destinations dependent on international tourists, with small populations less able to afford domestic leisure travel. The era of widespread, relatively cheap, international air travel that existed in the years before COVID-19 also supported greater mobility for work and study, resulting in many more families spread across the globe, and staycations are not a substitute for this distant travel required to reconnect with friends and family. The limited amount of academic attention paid to staycations is also of concern. There is considerable opportunity for academics to make a more active contribution to tourism sustainability by researching staycations and the leverage points listed in this chapter.

Despite these issues, there are positive signs that staycations are increasing in popularity in industry reports and commentary. The CEO of AirBnb reports that industry research emphasizes trends toward more meaningful travel, with travel and living becoming more closely linked (Brown, 2021). Surveys of UK residents also suggest an ongoing interest in staying closer to home for holidays and a return to traveler concerns about the sustainability of travel as we move to living with COVID-19 (Schofields Insurance, 2020). The key is to move to a new definition of a holiday as "simply a time when normal activity is suspended" (Swain, 2020, p. 1) and as an opportunity to improve one's quality of life by finding awe in the mundane.

References

Abson, D. J., Fischer, J., Leventon, J., Newig, J., Schomerus, T., Vilsmaier, U., von Wehrden, H., Abernethy, P., Ives, C. D., Jager, N. W., & Lang, D. J. (2017). Leverage points for sustainability transformation. *Ambio, 46*(1), 30–39. https://doi.org/10.1007/s13280-016-0800-y

Ali, R. (2020). *The winners & losers of a staycation world ahead.* SKIFT. https://skift.com/2020/05/06/the-winners-losers-of-a-staycation-world-ahead/

Arnold, R. D., & Wade, J. P. (2015). A definition of systems thinking: A systems approach. *Procedia Computer Science, 44*, 669–678. https://doi.org/10.1016/j.procs.2015.03.050

Besson, A. (2017). Everyday aesthetics on staycation as a pathway to restoration. *International Journal of Humanities and Cultural Studies, 4*(2), 34–52.

Betley, E., Sterling, E. J., Akabas, S., Paxton, A., & Frost, L. (2021). Introduction to systems and systems thinking. *Lessons in Conservation, 11*(1), 9–25.

Blank, C., Gatterer, K., Leichtfried, V., Pollhammer, D., Mair-Raggautz, M., Duschek, S., Humpeler, E., & Schobersberger, W. (2018). Short vacation improves stress-level and well-being in German-speaking middle-managers—A randomized controlled trial. *International Journal of Environmental Research and Public Health, 15*(1), 130. https://doi.org/10.3390/ijerph15010130

Bolchinova, A. (2020). *10 tourism marketing campaigns that work despite difficult times.* https://pro.regiondo.com/tourism-marketing-campaigns-covid/

Brouder, P. (2020). Reset redux: Possible evolutionary pathways towards the transformation of tourism in a COVID-19 world. *Tourism Geographies, 22*(3), 484–490. https://doi.org/10.1080/14616688.2020.1760928

Brown, V. (2021). Airbnb CEO Brian Chesky forecast on the future of global travel. *News.Com.* https://www.news.com.au/travel/travel-advice/accommodation/airbnb-ceo-brian-chesky-forecast-on-the-future-of-global-travel/news-story/0c3a0f4f85690f1fe58c5bf51b42725f

Cardini, B. B., & Freund, A. M. (2021). Recovery from accumulated strain: The role of daily mood and opportunity costs during a vacation. *Psychology & Health, 36*(8), 913–933. https://doi.org/10.1080/08870446.2020.1809661

Chen, C.-C., & Petrick, J. F. (2016). The roles of perceived travel benefits, importance, and constraints in predicting travel behavior. *Journal of Travel Research, 55*(4), 509–522. https://doi.org/10.1177/0047287514563986

de Bloom, J., Kompier, M., Geurts, S., De Weerth, C., Taris, T., & Sonnentag, S. (2009). Do we recover from vacation? *Journal of Occupational Health, 51*, 13–25.

de Bloom, J., Nawijn, J., Geurts, S., Kinnunen, U., & Korpela, K. (2017). Holiday travel, staycations, and subjective well-being. *Journal of Sustainable Tourism, 25*(4), 573–588. https://doi.org/10.1080/09669582.2016.1229323

Dinev, K. (2020). The return of domestic tourism. *IbisWorld*. https://skift.com/2020/05/06/the-winners-losers-of-a-staycation-world-ahead/

Draper, J. (2017). Staycation tourism. In L. Lowry (Ed.), *The SAGE international encyclopedia of travel and tourism* (pp. 1178–1179). Sage.

Edelheim, J., Molz, J. G., Jóhannesson, G. T., & Rantala, O. (2020). Slowing down with tourism. *Matkailututkimus, 16*(2), 4–8. https://doi.org/10.33351/mt.100493

Edgar, L. (2020). *Covid-19: Staycations key to supporting tourism sector.* https://www.theplanner.co.uk/news/covid-19-staycations-key-to-supporting-tourism-sector

Escape. (2021a). *16 places you won't believe are in Australia.* https://www.escape.com.au/destinations/australia/15-places-you-wont-believe-are-in-australia/image-gallery/7f86bcb8e5a724f93824d7e40f6c70c8

Escape. (2021b). *18 cool facts no one told me about Australia.* https://www.escape.com.au/top-lists/18-cool-facts-about-australia/image-gallery/dcb48a3920329ea364bb1bef600c25a6?galleryimage=10

Fox, S. (2009). Vacation or staycation? *Neumann Business Review.* https://citeseerx.ist.psu.edu/viewdoc/download?doi=10.1.1.474.7210&rep=rep1&type=pdf

Hall, C. M. (2011). Consumerism, tourism and voluntary simplicity. *Tourism Recreation Research, 36*(3), 298–303. https://doi.org/10.1080/02508281.2011.11081675

Herman, J., & Patel, D. (2018). Advising the traveller. *Medicine, 46*(1), 59–65. https://doi.org/10.1016/j.mpmed.2017.10.004

Inghels, D. (2020). *Introduction to modeling sustainable development in business processes.* Springer.

James, A., Ravichandran, S., Chuang, N. K., & Bolden, E., III. (2017). Using lifestyle analysis to develop lodging packages for staycation travelers. *Journal of Quality Assurance in Hospitality & Tourism, 18*(4), 387–415. https://doi.org/10.1080/1528008X.2016.1250240

Kastenholz, E., Eusébio, C., & Carneiro, M. J. (2018). Segmenting the rural tourist market by sustainable travel behaviour: Insights from village visitors in Portugal. *Journal of Destination Marketing & Management, 10*, 132–142. https://doi.org/10.1016/j.jdmm.2018.09.001

Keltner, D. (2016). Why do we feel Awe? *Greater Good.* Berkeley University of California. https://greatergood.berkeley.edu/article/item/why_do_we_feel_awe

Kessler, A., Pachucki, C., Stummer, K., Mair, M., & Binder, P. (2015). Types of organizational innovativeness and success in Austrian hotels. *International Journal of Contemporary Hospitality Management, 27*(7), 1707–1727. https://doi.org/10.1108/IJCHM-03-2014-0150

Kim, D. H. (1999). *Introduction to systems thinking.* https://thesystemsthinker.com/wp-content/uploads/2016/03/Introduction-to-Systems-Thinking-IMS013Epk.pdf

Knight, A. T., Cook, C. N., Redford, K. H., Biggs, D., Romero, C., Ortega-Argueta, A., et al. (2019). Improving conservation practice with principles and tools from systems thinking and evaluation. *Sustainability Science, 14*(6), 1531–1548. https://doi.org/10.1007/s11625-019-00676-x

Korzeniewski, K. (2020). Post-travel screening of symptomatic and asymptomatic travelers. *International Maritime Health, 71*(2), 129–139. https://doi.org/10.5603/IMH.2020.0023

KPMG. (2021). *Beyond Covid-19: Rise of domestic travel and tourism in Australia.* https://home.kpmg/au/en/home/insights/2020/12/beyond-covid-19-rise-of-domestic-travel-tourism-australia.html

Kuykendall, L., Craig, L., Stiksma, M., & Guarino, K. (2021). Understanding employees' unused vacation days. *Journal of Occupational Health Psychology, 26*(2), 69. https://doi.org/10.1037/ocp0000182

Lich, K. H., Urban, J. B., Frerichs, L., & Dave, G. (2017). Extending systems thinking in planning and evaluation using group concept mapping and system dynamics to tackle complex problems. *Evaluation and Program Planning, 60*, 254–264. https://doi.org/10.1016/j.evalprogplan.2016.10.008

Lin, Z. C., Wong, I. A., Kou, I. E., & Zhen, X. C. (2021). Inducing wellbeing through staycation programs in the midst of the COVID-19 crisis. *Tourism Management Perspectives, 40*. https://doi.org/10.1016/j.tmp.2021.100907

Long, I., & Flaherty, G. T. (2018). Traumatic travels—A review of accidental death and injury in international travellers. *International Journal of Travel Medicine and Global Health, 6*(2), 48–53. https://doi.org/10.15171/ijtmgh.2018.10

McCool, S., Freimund, W., & Breen, C. (2015). Benefiting from complexity thinking. In G. L. Worboys, M. Lockwood, A. Kothari, S. Feary, & I. Pulsford (Eds.), *Protected area governance and management* (pp. 291–326). ANU Press.

McCool, S. F., & Kline, J. D. (2020). A systems thinking approach for thinking and reflecting on sustainable recreation on public lands in an era of complexity, uncertainty, and change. In S. Selin, L. Cerveny, D. Blahna, & A. Miller (Eds.), *Igniting research for outdoor recreation: Linking science, policy, and action* (pp. 161–172). US Department of Agriculture, Forest Service, Pacific Northwest Research Station.

Mélon, M., Agrigoroaei, S., Diekmann, A., & Luminet, O. (2018). The holiday-related predictors of wellbeing in seniors. *Journal of Policy Research in Tourism, Leisure and Events, 10*(3), 221–240. https://doi.org/10.1080/19407963.2018.1470184

Molz, J. G. (2009). Representing pace in tourism mobilities. *Journal of Tourism and Cultural Change, 7*(4), 270–286. https://doi.org/10.1080/14766820903464242

Morgan, N. (2012). Time for 'mindful' destination management and marketing. *Journal of Destination Marketing & Management, 1*(1–2), 8–9. https://doi.org/10.1016/j.jdmm.2012.07.003

Moscardo, G. (2011). Exploring social representations of tourism planning: Issues for governance. *Journal of Sustainable Tourism, 19*(4–5), 423–436. https://doi.org/10.1080/09669582.2011.558625

Moscardo, G. (2019). Connecting people with experiences. In S. McCool & K. Bosak (Eds.), *A research agenda for sustainable tourism* (pp. 70–89). Edward Elgar.

Moscardo, G. (2020). Stories and design in tourism. *Annals of Tourism Research, 83*. https://doi.org/10.1016/j.annals.2020.102950

Moscardo, G. (2022). The responsible management of ecotourism experiences. In D. Agapito, M. Ribeiro, & K. Woosnam (Eds.), *Handbook on the tourist experience* (pp. 151–165). Edward Elgar.

Moscardo, G., & Murphy, L. (2016). Using destination community wellbeing to assess tourist markets. *Journal of Destination Marketing & Management, 5*(1), 55–64. https://doi.org/10.1016/j.jdmm.2016.01.003

Moya Calderón, M., Chavarría Esquivel, K., Arrieta García, M. M., & Lozano, C. B. (2021). Tourist behaviour and dynamics of domestic tourism in times of COVID-19. *Current Issues in Tourism*. https://doi.org/10.1080/13683500.2021.1947993

Parsons, A. (2021). Whitsunday eco-tourism offering a glimmer of hope for operators through pandemic. *ABC News*. https://www.abc.net.au/news/2021-08-29/eco-tourism-helps-businesses-through-pandemic/100405046

PeekPro. (2020). *The rise of staycations*. https://www.peek.com/pro/trends/the-rise-of-staycations

Ram, Y., Nawijn, J., & Peeters, P. M. (2013). Happiness and limits to sustainable tourism mobility. *Journal of Sustainable Tourism, 21*(7), 1017–1035. https://doi.org/10.1080/09669582.2013.826233

Reizer, A., & Mey-Raz, N. (2019). Slowing down vacation fade-out effects. *International Journal of Stress Management, 26*(3), 213–222. https://doi.org/10.1037/str0000103

Salmela, T., Nevala, H., Nousiainen, M., & Rantala, O. (2021). Proximity tourism. *Matkailututkimus, 17*(1), 46–63. https://doi.org/10.33351/mt.107997

Schofields Insurance. (2020). Report: The rise of staycations. https://www.schofields.ltd.uk/blog/5980/staycations-uk-travel-2020-21/

Smyth, A., De Bloom, J., Syrek, C., Domin, M., Janneck, M., Reins, J. A., & Lehr, D. (2020). Efficacy of a smartphone-based intervention—"Holidaily"-promoting recovery behaviour in workers after a vacation. *BMC Public Health, 20*(1), 1–17. https://doi.org/10.1186/s12889-020-09354-5

Sonnentag, S., Cheng, B. H., & Parker, S. L. (2022). Recovery from work. *Annual Review of Organizational Psychology and Organizational Behavior, 9*, 33–60. https://doi.org/10.1146/annurev-orgpsych-012420-091355

Sonnentag, S., & Niessen, C. (2020). To detach or not to detach? *Frontiers in Psychology, 11*. https://doi.org/10.3389/fpsyg.2020.560156

Sørensen, E. B., & Hjalager, A.-M. (2020). Conspicuous non-consumption in tourism: Non-innovation or the innovation of nothing? *Tourist Studies, 20*(2), 222–247. https://doi.org/10.1177/1468797619894463

Sudevan, P. (2020). How staycations are becoming the most popular tourism trend after COVID-19. *The Hindu News*. https://www.thehindu.com/life-and-style/travel/how-staycations-are-becoming-the-most-popular-tourism-trend-after-covid-19/article31963261.ece

Swain, S. (2020). A message for anyone who thinks traveling in their own country is boring. *Nine News*. https://travel.nine.com.au/latest/tasmania-road-trips-hobart-australia-holidays-coronavirus/6bf58a88-c5b1-43a6-acad-ca903d949e6c

UNWTO. (2020). *Understanding domestic tourism and seizing its opportunities*. UNWTO. https://www.e-unwto.org/doi/pdf/10.18111/9789284422111

Vogt, C., & Andereck, K. (2018). Are we there yet on sustainable tourism? *Journal of Park and Recreation Administration, 36*, 212–214.
Wei, C., Zhao, W., Zhang, C., & Huang, K. (2019). Psychological factors affecting memorable tourism experiences. *Asia Pacific Journal of Tourism Research, 24*(7), 619–632. https://doi.org/10.1080/10941665.2019.1611611
West, C., Mogilner, C., & DeVoe, S. E. (2021). Happiness from treating the weekend like a vacation. *Social Psychological and Personality Science, 12*(3), 346–356. https://doi.org/10.1177/1948550620916080
White, G. (2021). BURNOUT: Helping employees avoid and weather the challenges of working from home. *Professional Safety, 66*(5), 16–19.

7

Local Tourist Shopping: How Purchases in a Destination Can Contribute to Sustainability

Alba García-Milon and Emma Juaneda-Ayensa

Introduction

Prior to the COVID-19 pandemic, the travel and tourism sector accounted for one in four new jobs created, 10.6% of total global jobs (334 million), 10.4% of the global GDP ($9.2 trillion), and $1.7 trillion was spent by international visitors in 2019 (6.8% of total exports, 27.4% of global services exports) (World Travel & Tourism Council, 2021). In 2019, these figures suggested that the tourism sector was entering its golden age, with forecasts projecting continued growth to rival the record-breaking trend experienced during the second half of the 1920s. However, the outbreak of the SARS-CoV-2 (COVID-19) virus at the end of 2019 paralyzed the lives of the global populations and tourism, as an economic entity, suffered the most during the pandemic recording a loss of almost $4.5 billion, a decline of 62 million jobs, and a 45%

A. García-Milon (✉) • E. Juaneda-Ayensa
University of La Rioja, Logroño, Spain
e-mail: alba.garciam@unirioja.es; emma.juaneda@unirioja.es

T. Walker et al. (eds.), *Sustainable Tourism*,
https://doi.org/10.1007/978-3-031-43528-7_7

decrease in spending by domestic visitors and of 69.4% by international visitors (World Travel & Tourism Council, 2021).

This period has allowed us to reflect on the influence of some tourism activities and how we should approach future development models. The Organisation for Economic Co-operation and Development (OECD, 2020) notes that responding to the COVID-19 crisis requires mutually supportive and integrated policies between economic, social, and environmental objectives to address the deeper nature of the crisis. Addressing the underlying causes of vulnerability in our systems, which stem from inequalities, injustice and discrimination, weak governance and institutions, inadequate public services and infrastructure, natural resource depletion, and biodiversity and climate degradation, will be essential in building resilience against future outbreaks. In this context, the United Nation's Sustainable Development Goals (SDGs) take on even greater relevance today, as they aim to address precisely these vulnerabilities and can be used as a framework to guide recovery from the COVID-19 crisis.

In relation to the tourism sector and its recovery, there are many questions about how we should rethink the tourism model of certain territories, considering that factors such as climate change or the fuel crisis will have an influence on how this model is to be defined. The concept of sustainable tourism, however, is not new. According to the definition given by the World Tourism Organization in 2005, sustainable tourism development meets the needs of present tourists and can mitigate the destruction of the environment and the seasonal nature of work while protecting and enhancing opportunities for the future (UNWTO, 2005). Sustainable tourism has gained popularity during the last decades due, for the most part, to the negative impact that unstructured and overcrowded tourism has had. For this reason, in 2005, the World Tourism Organization and the United Nations Environment Program established 12 objectives that constitute a model for sustainable tourism (Sustainable Tourism for Development, 2013, p. 18):

1. Economic viability: To ensure the viability and competitiveness of tourism destinations and enterprises, so that they are able to continue to prosper and deliver benefits in the long term.

2. Local prosperity: To maximize the contribution of tourism to the prosperity of the host destination, including the proportion of visitor spending that is retained locally.
3. Employment quality: To strengthen the number and quality of local jobs created and supported by tourism, including the level of pay, conditions of service, and availability to all without discrimination by gender, race, disability, or other ways.
4. Social equity: To seek a widespread distribution of economic and social benefits from tourism throughout the recipient community, including improving opportunities, income, and services available to the poor.
5. Visitor fulfillment: To provide a safe, satisfying, and fulfilling experience for visitors, available to all without discrimination by gender, race, disability, or other ways.
6. Local control: To engage and empower local communities in planning and decision-making about the management and future development of tourism in their area, in consultation with other stakeholders.
7. Community wellbeing: To maintain and strengthen the quality of life in local communities, including social structures and access to resources, amenities, and life support systems, avoiding any form of social degradation or exploitation.
8. Cultural richness: To respect and enhance the historic heritage, authentic culture, traditions, and distinctiveness of host communities.
9. Physical integrity: To maintain and enhance the quality of landscapes, both urban and rural, and avoid the physical and visual degradation of the environment.
10. Biological diversity: To support the conservation of natural areas, habitats, and wildlife and minimize damage to them.
11. Resource efficiency: To minimize the use of scarce and non-renewable resources in the development and operation of tourism facilities and services.
12. Environmental purity: To minimize the pollution of air, water, and land and the generation of waste by tourism enterprises and visitors.

In essence, compliance with this model is expected to lead to the management of all resources in such a way that economic, social, and environmental needs can be fulfilled while maintaining cultural integrity, essential ecological processes, biological diversity, and life support systems.

Although sustainable tourism is a widely used concept, its practical application and evaluation through indicator systems to ensure sustainability standards is still a work in progress (Asmelash & Kumar, 2019). To avoid "green washing," where sustainability is implied but seldom integrated as a guiding principle and implies a problem for the management of tourist destinations and their social and environmental consequences, sustainable tourism should generate measurable impacts in income for local communities and contribute to their economic, social, and cultural development and wellbeing, and in developing countries, tourism should contribute significantly to poverty reduction (Croes, 2014).

In this context, shopping is a common and fundamental tourist activity both from the tourist's perspective and for the chosen destination (Jin et al., 2017). This act is not only utilitarian with a focus on acquiring daily necessities (Timothy, 2005) but also a way to obtain souvenirs of the travel experience, understand the local culture, and strengthen social networks (García-Milon et al., 2020). Destinations have the opportunity to offer visiting tourists the chance to enjoy local products on-site and thus enhance the development of the region's economy. Therefore, the proposal that tourist destination managers should make to generate value for their visitors should be based on the offer of products or services that are specific to the area, linked to its natural and cultural resources, which give uniqueness to the experience and differentiate it from other tourist destinations. This added value is one that must be built collaboratively among different agents and stakeholders in which each of the stakeholders provides value and ensures uniqueness to make the experience difficult to imitate and therefore enhance the competitiveness of the destination. Although tourist shopping has recently attracted a lot of attention from researchers, its role in the development of sustainable tourism has not yet been analyzed (Jin et al., 2017). This chapter begins to address this gap by establishing the benefits of shopping for sustainable tourism and defines a new practice that links tourist shopping and sustainability: local tourist shopping.

The objective of this chapter is to present the importance of local tourist shopping in promoting the sustainability of tourist destinations. The first section explores consumerism and its connection to sustainability. The second section concisely presents the relationship between tourism purchasing and sustainable tourism and the main agents involved. The third section introduces the concept of local tourist shopping, how it results in a sustainable tourism practice, as well as the stakeholders involved in the activity; finally, the last section presents a set of conclusions and recommendations to promote this activity and become a local shopping destination based on sustainability.

Consumerism and Sustainability

Consumer habits and behaviors are constantly changing. Increasingly, sustainability in products and services is more valued and positively considered when purchasing. This has led to a new trend referred to as sustainable consumerism which goes beyond the consumption of sustainable products as it implies considering all activities involved in the production process as well as lifestyle practices, familiar behavior, and technology (Lim, 2017; Liu et al., 2016) in order to mitigate negative effects to health, quality of living, environment, and consumers (Perera et al., 2018; Pocol et al., 2020).

Sustainable consumer behavior has traditionally been associated with environmental awareness. In fact, it was defined by White et al. (2019) as "actions that result in decreases of adverse environmental impacts as well as decreased utilization of natural resources across the lifecycle of the product." However, this concept is wider and goes beyond the environment. The social aspect is an essential pillar in sustainability and can be a key element in the purchase intention of individuals, giving rise to concepts such as pro-social consumers. These are private individuals devoted to avoiding products that apparently harm society. In order to reduce their negative impact and maximize benefits to society, they deliberately purchase pro-social products and conduct boycotts to those which present adverse effects (Berki-Kiss & Menrad, 2022).

Sustainable consumerism supports certain social issues: labor maintenance and generation as well as access to fundamental rights such as security, health, and education, amongst others (Pocol et al., 2020). The social impact of purchases, such as labor conditions, impact on living conditions, the welfare of growers and producers, or child labor (Fernando et al., 2018) are becoming increasingly important to consumers (Baskentli et al., 2019). They have transformed sustainability into a business imperative that must be followed by companies and retailers. To do so, the introduction of related certifications has been promoted to comply with consumers increasing demands and socially responsible expectations (Emberger-Klein & Menrad, 2018; Van Loo et al., 2014).

Tourists also have an increased awareness and look to travel to sustainable destinations. Although there is a general idea of consumerism associated with tourism, destinations can increase their sustainability by incorporating activities and practices that have a positive impact on locals and the physical destination, attracting pro-social tourists. Previous studies have found that sustainable tourism policies can improve tourist destination success where the use of eco-brands, eco-labels, and environmental ads has a positive influence on tourists' shopping behavior (Chin et al., 2018).

Sustainable Management in Tourist Destinations

The United Nations Environment Program (UNEP) and United Nations World Tourism Organization (UNWTO) (2005) define sustainable tourism as "tourism that takes full account of its current and future economic, social and environmental impacts, addressing the needs of visitors, the industry, the environment, and host communities". Sustainable tourism is understood as a three-dimensional concept including economic, sociocultural, and environmental sustainability (Mohamadi et al., 2022; Ritchie & Crouch, 2003). Therefore, it should include practices in the long-term that ensure the maintenance and improvement of a wide variety of aspects from these three dimensions. Specifically, natural, social, and cultural diversity, the development of tourism in a framework of local and national strategic planning, the support of the local economic system, participation in local communities, consultancy with stakeholders and locals, and the monitoring of progressing studies through the industry by analyzing and collecting data effectively comprise these sustainable tourism principles (Goodwin, 2011; Mohamadi et al., 2022).

The maintenance of communities' diversity and support of local economies are then key to the development of sustainable tourism. Communities can benefit from tourism to sustain and evolve, avoiding a loss of identity. Tourism has always been considered appropriate for reducing poverty and boosting economies (Croes, 2014). Among other forms of economic stimuli, tourism can be employed as a growth engine to reduce poverty in traditional communities (Croes, 2014; Lee & Jan, 2019). Tourism can create employment for residents (UNWTO, 2005) and generate opportunities to sell local products (Muro-Rodríguez et al., 2020). As tourists consume a great number of products and services at their destinations, many chances emerge for host economies (Croes, 2014): higher employment opportunities and higher income levels can have a trickle-down effect. Consequently, destinations should be able to sustain themselves thanks to the development of sustainable tourism. However, it is worth highlighting that sustainable tourism requires a holistic and systemic approach that allows individuals, businesses, and organizations to integrate sustainable practices in their daily operations and functions (Roxas et al., 2020).

This is of utmost importance as stakeholders in the sector are at the core of sustainable tourism. There are many stakeholders identified in sustainable tourism (Lee & Jan, 2019) such as visitors, residents, for-profit organizations, government entities, non-profit organizations, and the environment. They offer views on how to manage tourism destinations that adhere to the economic, environmental, and socio-cultural objectives of sustainable tourism, as well as to jointly address sustainability issues facing the sector (Jamal & Stronza, 2009; Roxas et al., 2020). Stakeholder collaboration has been seen as a major factor in the pursuit of sustainable tourism (Jamal & Stronza, 2009; Wondirad et al., 2020).

Concerns have grown, in the field of sustainability management, around the concept of "green washing", a term popularized in the mid-1990s. Its narrower application relates the concept to exclusively environmental issues (e.g. Chen & Chang, 2013; Pearson, 2010). However, considering the principles of environmental integrity, social equity, and economic prosperity, the term can speak to environmental, social, and economic issues (Seele & Gatti, 2017).

Local Tourist Shopping

Shopping is an important activity in tourism. It has great benefits for destinations as it can contribute to a positive image and generate a significant amount of revenue (García-Milon et al., 2021a; Jin et al., 2017). In addition, shopping can be an element of satisfaction and pleasure for tourists and an enhancement to the tourist experience (Hsieh & Chang, 2006; Meng et al., 2019; Way & Robertson, 2013). Therefore, this activity has been studied and promoted by academics and practitioners in the field of tourism (García-Milon et al., 2021a). Indeed, the term tourist shopping has been defined as a recreational activity that complements the travel where tourists search for, select, and buy goods in a destination (García-Milon et al., 2021c; Jin et al., 2017). This is an extensive concept where any good or product can be included regardless of its origin or typology and all product purchases in a destination can be included in the activity. The complete process of purchasing in a destination by tourists is called the tourist shopping journey (TSJ) (García-Milon et al., 2020).

As we dive deeper into this concept, questions surface regarding the importance of the value of an object purchased in a specific destination. There might be different answers to this question, and different aspects related to price, convenience, and impulsive buying can also arise. However, another significant reason to purchase products in a destination could be motivated by a combination of the product and the destination that together create the unique experience (García-Milon et al., 2021b). In other words, tourist shopping can become inimitable and priceless if local cultures, local resources, and local people are a vital part of the purchased product. This results in a new concept, coined "local tourist shopping" as a particular activity within the notion of tourist shopping.

Local tourist shopping can be defined as a purchasing activity undertaken by tourists in which the acquired products are locally produced and distributed and are significant to the destination's culture and traditions. When undertaking this activity, concepts such as KM0, traditional manufacturing, or craftwork can be crucial in the decision to purchase products at the destination. Also, local tourist shopping should involve locals in a direct interaction with tourists in the role of producers, manufacturers, or sellers. This activity materializes intangible assets such as uniqueness, tradition, and culture in tangible products that can be transported to tourists' country of origin. Additionally, the traditional products offered to tourists in this activity are generated employing autochthonous natural resources and indigenous knowledge. Therefore, local tourist shopping has significant benefits for both tourists and destinations; Figure 7.1 illustrates how these different spheres interact. On the one hand, tourists can acquire a product that captures a destination's

Fig. 7.1 Interrelation of benefits in local tourist shopping. (Source: author's own)

culture and tradition, allowing them to experience local culture in a different manner. On the other hand, destinations can share their customs and national identities by offering local products to tourists while generating earnings for the local population.

Furthermore, this activity can be an appreciated contribution to sustainable development and lead to the creation of sustainable destinations. Regarding the Sustainable Development Goals (UNWTO, 2022), local tourist shopping can be key to achieving goals number 8 and 12: "decent work and economic growth" and "ensure sustainable consumption and production patterns." Specifically, target 8.9 recognizes the tourism sector's contribution to job creation which can increase skills and professional development. With regard to goal 12, target 12.b aims to develop sustainable tourism able to create jobs and promote local culture and products. Furthermore, the UNWTO developed a platform that showed the potential of tourism to contribute to all of the Sustainable Development Goals (UNWTO, 2022). Table 7.1 shows how the Sustainable Development Goals can be also benefited from local tourist shopping.

In addition, Asmelash and Kumar (2019) developed a set of indicators to assess the sustainability of tourism. Some of them are closely related to the consecution of local tourist shopping. Therefore, this strengthens the value of local-based tourist shopping in the path to becoming a sustainable tourism destination. Table 7.2 presents the indicators that could be impacted by local tourist tourism shopping.

Local tourist shopping is a complete activity in the pursuit of sustainable tourism since it is able to benefit the three dimensions that converge in the term (Mohamadi et al., 2022; Ritchie & Crouch, 2003) as well as institutional sustainability. This activity creates an ecosystem that integrates locals and residents as a key element to generate unique physical products and experiences for tourists. Therefore, to avoid "green washing" in this activity, it is of utmost importance to ensure that the products offered and their production follow the principles of environmental integrity, social equity, economic prosperity, and revert to locals. For the purpose of our analysis, we define the main groups to include in our evaluation of the impacts of local tourist shopping as being part of the four following categories:

Table 7.1 Local tourist shopping and the SDGs

1 NO POVERTY	Provide income through job creation, promoting entrepreneurship, and small businesses related to the production and distribution of local products	10 REDUCED INEQUALITIES	Local tourist shopping engages local populations and key stakeholders. It can contribute to urban renewal and rural development by giving people the opportunity to prosper locally
2 ZERO HUNGER	Sustainable agriculture and local products can generate additional income while enhancing the value of the tourism experience	11 SUSTAINABLE CITIES AND COMMUNITIES	Local tourist shopping can promote regeneration and preserve cultural and natural heritage by protecting local products
3 GOOD HEALTH AND WELL-BEING	Tax income from local products purchased by tourists can be reinvested in health care and services	12 RESPONSIBLE CONSUMPTION AND PRODUCTION	Sustainable consumption and production (SCP) are promoted by local tourist shopping as it employs local resources and highlights the value of the production process
4 QUALITY EDUCATION	A skilled workforce is crucial for local tourist shopping to prosper. The activity provides opportunities for direct and indirect jobs	13 CLIMATE ACTION	Local tourist shopping is environmentally sustainable as it promotes the responsible use of local natural resources and prioritizes environmentally friendly products
5 GENDER EQUALITY	The provision of direct jobs and income generation from SMEs in local tourist shopping can empower women	14 LIFE BELOW WATER	The activity can serve as a vehicle to promote a blue economy, contributing to the sustainable use of marine resources
6 CLEAN WATER AND SANITATION	The production of local products encourages efficient use of water, pollution control, and technology efficiency	15 LIFE ON LAND	Rich biodiversity and natural heritage can be key to attractive products for tourists whose purchases can generate revenue for local communities and the preservation of nature

(*continued*)

Table 7.1 (continued)

7	Local tourist shopping can accelerate the shift toward increased renewable energy by promoting investments in clean production processes	16 PEACE, JUSTICE AND STRONG INSTITUTIONS	Local tourist shopping can foster multicultural and inter-faith tolerance and understanding. It engages local communities and can also contribute to peace in post-conflict societies
8 DECENT WORK AND ECONOMIC GROWTH	Local tourist shopping can promote decent work opportunities and policies that favor better diversification	17 PARTNERSHIPS FOR THE GOALS	The activity can strengthen private/public partnerships and engage multiple stakeholders to work together to achieve the SDGs
9 INDUSTRY, INNOVATION AND INFRASTRUCTURE	Local tourist shopping relies on good public and private infrastructure, especially high streets and stores. The activity can influence infrastructure upgrade, thus attracting tourists		

Source: Adapted from UNWTO (2022)

Table 7.2 Indicators of sustainability in tourism

Economic sustainability	Employment quality	• The number of job opportunities for the local residents • Percentage of quality (stable, high-paid, permanent, and full-time) tourism jobs
	Economic viability	• Level of local economic diversification due to heritage tourism • Amount of income to the local communities
	Local prosperity	• Variety of local products available because of heritage tourism • Availability of markets for local products • Degree of financial leakage away from the destination
Environmental sustainability	Physical integrity	• Integration of heritage tourism and the environment • Suitability of facilities to heritage tourism
Socio-cultural sustainability	Visitor fulfilment	• Percentage of tourists encouraged to learn about local cultures • Quality of host-guest interaction
	Local control	• Local residents' knowledge of heritage tourism and its sustainability • Presence of local help for residents on how to portray their culture to tourists
	Community wellbeing	• Contribution of local cultural values to heritage tourism development • Retention of local lifestyles • Quality of recreational opportunities for residents due to heritage tourism
Institutional sustainability	Local-oriented control policy	• Controlling practices of local tourism development • Presence of tourism planner among the local residents • Inclusion of indigenous knowledge in heritage sites management

Source: Adapted from Asmelash and Kumar (2019)

Tourists: Regardless of their different motivations, all seek to have experiences that allow them to create positive memories which will allow them to value their trip. Memorable Tourism Experiences (MTEs) can activate long-lasting memories. This helps in understanding that it is not the type of experience itself that makes a trip memorable, but rather how each experience represents a meaningful moment in the consumer's life (Rasoolimanesh et al., 2021). Three dimensions appear to be key to memorable experiences and influence the memory arousal of travel experiences: the personal/psychological, the relational, and the environmental/cultural. When shopping for a product in a destination, the tourist's choice not only lies in the physical attributes of the product but also in other attributes related to local culture and strengthening social bonds (Muro-Rodríguez et al., 2020; Timothy, 2005).

Governments: The nature and structure of the retail labor force make the sector important with regard to employment policies. The increased popularity of digital commerce has abolished many jobs in the sector, mostly in small local stores (Whysall, 2000). Therefore, in the public sector, policies are defined to support small retailers because the income of entire families depend on these jobs, but also because the incomes of the owners and their subsidiary companies do as well. On the other hand, small businesses contribute through the generation of income that stays in the territory and is channeled through the payment of taxes that result in the improvement of public revenues.

Retailers: Local retailers are the first to be able to offer their products as something different and generate value for customers, in this case, tourists. For them, tourism can be an opportunity in terms of market segment. They can contribute to offering a service that goes beyond the availability of local products as they can offer a unique shopping experience that contributes to making valuable memories. Tourists' evaluations of these experiences are an attraction for future visitors who may visit the establishments for the experience itself, beyond the product they purchase. There has been a recent increase in the number of travelers choosing their destinations based on sustainability aspects. The relevance of this social aspect is often neglected, although it constitutes a fundamental pillar of the concept of sustainability and is also assumed to guide the purchase intention of individuals. Some recent studies confirm that sustainability-oriented consumers are more inclined to choose these types of products when social and environmental benefits are actively promoted (White et al., 2019).

Residents: Local commerce gives life to cities and municipalities, not only because it creates activity around its establishments, but also because it favors the promotion of local culture. The streets and public spaces are direct beneficiaries of the activity that takes place in the stores, favoring active coexistence spaces for citizens.

Discussion and Recommendations

There is a global need to identify and promote sustainable tourism practices. Shopping at destinations can be a successful manner to achieve this goal when managed and handled properly. Therefore, it must involve traditional and local retail stores and autochthonous products so that purchases have a direct impact on the destination and its locals.

Local tourist shopping is in line with the consecution of tourism-related Sustainable Development Goals (UNWTO, 2022) such as goals 8 and 12 and can help accomplish some of the indicators for sustainability in tourism (Asmelash & Kumar, 2019) since it is able to benefit the three dimensions that converge in the term (i.e., economic, social-cultural, and environmental sustainability) (Mohamadi et al., 2022; Ritchie & Crouch, 2003) and also institutional sustainability.

In order to boost this activity and to create a local shopping destination based on sustainability, we outline five recommendations in Fig. 7.2. All of them present the value created along the local tourist shopping journey:

Recommendation 1 (R1). Networking: Interconnect the stakeholders at destinations, promoting mutual support between retailers, producers, and destination managers. Stakeholders and their cooperation are essential in making a destination sustainable. In local tourist shopping, the agents involved in the production and selling of products must be visible parts of the destination's offerings. The main goal is to build synergies and positive effects to add value to the tourist destination.

Fig. 7.2 Recommendations for local tourist shopping. (Source: author originated)

Recommendation 2 (R2). Catalogue of local and autochthonous products: Identifying indigenous products is the first step in protecting and promoting the resources and techniques involved in their creation. In this sense, the elaboration of a catalogue of local items facilitates the identification of elements that represent the destination culture. At the same time, it contributes to territory protection as this measure impacts the whole production process and the activities involved. It also supports the natural resources and jobs that are traditionally important in the region.

This identification and classification in a catalogue of local products would facilitate proper communication channels to promote local producers and retailers' products abroad, incorporating them into destination advertising. This would give visibility and voice on an international scale to local tourist shopping and spread the concept's benefit (social, economic, and environmental) so tourists can be interested in purchasing local products as a sustainable practice at the destination.

It would be imperative, in the distribution of such a catalogue, to prioritize quality over quantity with regard to the promotion of destination shopping. Such a catalogue would also need to highlight the direct benefits of purchasing local products over globalized goods. By communicating the positive impacts of purchasing local products and highlighting the indigenous nature of materials, production process, and workers involved in its production, a catalogue could attract pro-social consumers and tourists, thus increasing the destination's sustainability.

Recommendation 3 (R3). Certification: Develop measures to emphasize the value of local products by indicating the autochthonous materials for production, their manufacturing process using traditional techniques, the applications and meanings of the products within the local culture, and eco-friendly production with official labels and certificates. This is imperative to ensure, in an official manner, that the products are locally sourced and produced and that the production process is sustainable.

Recommendation 4 (R4). Tourists' experiences: Create tourist experiences through the elaboration of storytelling for local products. This might create an emotional connection between tourists and the host destination, generating an interest in taking part in the protection of tradition and culture by purchasing a part of it: local products. In this way, tourists acquire more than a souvenir, they expand their tourism experience and contribute to the maintenance of a destination's identity.

In the creation of experiences, it could also be generative to imagine the design of guided tours, where local products are included as an important element of culture and tradition allowing tourists to interact with local producers. This connects tourists and locals so they can share their culture.

Recommendation 5 (R5). Logistic service: Combine efforts and promote the collaboration of manufacturers or traders to offer joint shipping services to the tourists' country of origin for products categorized as local, especially in those cases where items are voluminous. This is important in the case of international tourists, as carrying the items home can be a barrier to purchasing products.

Conclusion

Tourism is a key sector of activity which can generate income for a destination, especially relevant for those developing countries with a valuable cultural and natural heritage. However, concerns have been raised that call for fairer behavior in the industry, as both negative impacts on the environmental sphere and the unbalanced redistribution of the gross income generated by tourism activity have been denounced. It is therefore important to clearly identify good practices linked to tourism so that a sustainable tourism model can be built. One of the practices linked to the objectives set for sustainable tourism is local tourist shopping, as a result of the great benefits to the destination. However, it is important to undertake practices that ensure that the activity is completely aligned with sustainable goals and the underlying message of holistic sustainability is correctly transmitted to tourists. In this sense, the cooperation of all the stakeholders involved is crucial. Destinations must protect and promote local tourist shopping as a manner to spread and conserve local culture, protect local resources, and enfranchise local people while providing a unique experience to tourists.

References

Asmelash, A. G., & Kumar, S. (2019). Assessing progress of tourism sustainability: Developing and validating sustainability indicators. *Tourism Management, 71*, 67–83. https://doi.org/10.1016/j.tourman.2018.09.020

Baskentli, S., Sen, S., Du, S., & Bhattacharya, C. (2019). Consumer reactions to corporate social responsibility: The role of CSR domains. *Journal of Business Research, 95*, 502–513. https://doi.org/10.1016/j.jbusres.2018.07.046

Berki-Kiss, D., & Menrad, K. (2022). The role emotions play in consumer intentions to make pro-social purchases in Germany–an augmented theory of planned behavior model. *Sustainable Production and Consumption, 29*, 79–89. https://doi.org/10.1016/j.spc.2021.09.026

Chen, Y.-S., & Chang, C.-H. (2013). Greenwash and green trust: The mediation effects of green consumer confusion and green perceived risk. *Journal of Business Ethics, 114*(3), 489–500. http://www.jstor.org/stable/23433794

Chin, C.-H., Chin, C.-L., & Wong, W. P.-M. (2018). The implementation of green marketing tools in rural tourism: The readiness of tourists? *Journal of Hospitality Marketing & Management, 27*(3), 261–280. https://doi.org/10.1080/19368623.2017.1359723

Croes, R. (2014). The role of tourism in poverty reduction: An empirical assessment. *Tourism Economics, 20*(2), 207–226. https://doi.org/10.5367/te.2013.0275

Emberger-Klein, A., & Menrad, K. (2018). The effect of information provision on supermarket consumers' use of and preferences for carbon labels in Germany. *Journal of Cleaner Production, 172*, 253–263. https://doi.org/10.1016/j.jclepro.2017.10.105

Fernando, A. L., Rettenmaier, N., Soldatos, P., & Panoutsou, C. (2018). Sustainability of perennial crops production for bioenergy and bioproducts. In *Perennial grasses for bioenergy and bioproducts* (pp. 245–283). Academic Press.

García-Milon, A., Juaneda-Ayensa, E., Olarte-Pascual, C., & Pelegrín-Borondo, J. (2020). Towards the smart tourism destination: Key factors in information source use on the tourist shopping journey. *Tourism Management Perspectives, 36*. https://doi.org/10.1016/j.tmp.2020.100730

García-Milon, A., Olarte-Pascual, C., & Juaneda-Ayensa, E. (2021a). Assessing the moderating effect of COVID-19 on intention to use smartphones on the tourist shopping journey. *Tourism Management, 87*. https://doi.org/10.1016/j.tourman.2021.104361

García-Milon, A., Olarte-Pascual, C., Juaneda-Ayensa, E., & Pelegrín-Borondo, J. (2021b). Tourist purchases in a destination: What leads them to seek information from digital sources? *European Journal of Management and Business Economics, 30*(2), 243–260. https://doi.org/10.1108/EJMBE-09-2019-0153

García-Milon, A., Pelegrín-Borondo, J., Juaneda-Ayensa, E., & Olarte-Pascual, C. (2021c). The smartphone: The tourist's on-site shopping friend. An extended cognitive, affective, normative model. *Telematics and Informatics, 61*. https://doi.org/10.1016/j.tele.2021.101618

Goodwin, H. (2011). *Taking responsibility for tourism: Responsible tourism management*. Goodfellow Publishers.

Hsieh, A.-T., & Chang, J. (2006). Shopping and tourist night markets in Taiwan. *Tourism Management, 27*(1), 138–145. https://doi.org/10.1016/j.tourman.2004.06.017

Jamal, T., & Stronza, A. (2009). Collaboration theory and tourism practice in protected areas: Stakeholders, structuring and sustainability. *Journal of Sustainable Tourism, 17*(2), 169–189. https://doi.org/10.1080/09669580802495741

Jin, H., Moscardo, G., & Murphy, L. (2017). Making sense of tourist shopping research: A critical review. *Tourism Management, 62*, 120–134. https://doi.org/10.1016/j.tourman.2017.03.027

Lee, T. H., & Jan, F.-H. (2019). Can community-based tourism contribute to sustainable development? Evidence from residents' perceptions of the sustainability. *Tourism Management, 70*, 368–380. https://doi.org/10.1016/j.tourman.2018.09.003

Lim, W. M. (2017). Inside the sustainable consumption theoretical toolbox: Critical concepts for sustainability, consumption, and marketing. *Journal of Business Research, 78*, 69–80. https://doi.org/10.1016/j.jbusres.2017.05.001

Liu, W., Oosterveer, P., & Spaargaren, G. (2016). Promoting sustainable consumption in China: A conceptual framework and research review. *Journal of Cleaner Production, 134*, 13–21. https://doi.org/10.1016/j.jclepro.2015.10.124

Meng, F., Zhang, P., Li, H., & So, K. K. F. (2019). Modeling precursors of impulsive tourist shopping behavior: Evidence from long-haul Chinese outbound tourists. *International Journal of Tourism Research, 21*(3), 344–358. https://doi.org/10.1002/jtr.2266

Mohamadi, S., Abbasi, A., Ranaei Kordshouli, H. A., & Askarifar, K. (2022). Conceptualizing sustainable–responsible tourism indicators: An interpretive structural modeling approach. *Environment, Development and Sustainability, 24*(1), 399–425. https://doi.org/10.1007/s10668-021-01442-9

Muro-Rodríguez, A. I., Pérez-Jiménez, I. R., & Sánchez-Araque, J. A. (2020). Impact of shopping tourism for the retail trade as a strategy for the local development of cities. *Frontiers in Psychology, 11*, 1–5. https://doi.org/10.3389/fpsyg.2020.00067

OECD. (2020, November). *Building a coherent response for a sustainable post-COVID-19 recovery* (pp. 1–28). OECD Publishing. https://legalinstruments.oecd.org/en/instruments/OECD-LEGAL-0381

Pearson, J. (2010). Turning point. Are we doing the right thing? Leadership and prioritisation for public benefit. *Journal of Corporate Citizenship, 37*, 37–40. http://www.jstor.org/stable/jcorpciti.37.37

Perera, C., Auger, P., & Klein, J. (2018). Green consumption practices among young environmentalists: A practice theory perspective. *Journal of Business Ethics, 152*(3), 843–864. https://doi.org/10.1007/s10551-016-3376-3

Pocol, C. B., Marinescu, V., Amuza, A., Cadar, R.-L, & Rodideal, A. A. (2020). Sustainable vs. unsustainable food consumption behaviour: A study among students from Romania. https://kopernio.com/viewer?doi=10.3390/su12114699&route=6

Rasoolimanesh, S. M., Seyfi, S., Hall, C. M., & Hatamifar, P. (2021). Understanding memorable tourism experiences and behavioural intentions of heritage tourists. *Journal of Destination Marketing & Management, 21*, 100621. https://doi.org/10.1016/j.jdmm.2021.100621

Ritchie, J. B., & Crouch, G. I. (2003). *The competitive destination: A sustainable tourism perspective*. Cabi.

Roxas, F. M. Y., Rivera, J. P. R., & Gutierrez, E. L. M. (2020). Framework for creating sustainable tourism using systems thinking. *Current Issues in Tourism, 23*(3), 280–296. https://doi.org/10.1080/13683500.2018.1534805

Seele, P., & Gatti, L. (2017). Greenwashing Revisited: In Search for a Typology and Accusation-based Definition Incorporating Legitimacy Strategies. *Business Strategy and the Environment, 26*(2), 239–252. https://doi.org/10.1002/bse.1912

Sustainable Tourism for Development. (2013). *Sustainable tourism for development guidebook* (1st ed). https://www.e-unwto.org/doi/pdf/10.18111/9789284415496

Timothy, D. J. (2005). *Shopping tourism, retailing, and Leisure*. Channel View Publications. https://books.google.es/books/about/Shopping_tourism_retailing_and_leisure.html?id=dfAJAQAAMAAJ&redir_esc=y

UNWTO. (2005). *Making tourism more sustainable—A guide for policy makers* (pp. 11–12). https://www.unwto.org/sustainable-development

UNWTO. (2022). *Tourism & sustainable development goals—Tourism for SDGs*. https://tourism4sdgs.org/tourism-for-sdgs/tourism-and-sdgs/

Van Loo, E. J., Caputo, V., Nayga, R. M., Jr., & Verbeke, W. (2014). Consumers' valuation of sustainability labels on meat. *Food Policy, 49*, 137–150. https://doi.org/10.1016/j.foodpol.2014.07.002

Way, K. A., & Robertson, L. J. (2013). Shopping and tourism patterns of attendees of the bikes, blues & BBQ festival. *Journal of Hospitality Marketing and Management, 22*(1), 116–133. https://doi.org/10.1080/19368623.2012.627261

White, K., Habib, R., & Hardisty, D. J. (2019). How to SHIFT consumer behaviors to be more sustainable: A literature review and guiding framework. *Journal of Marketing, 83*(3), 22–49. https://doi.org/10.1177/0022242919825649

Whysall, P. (2000). Addressing ethical issues in retailing: A stakeholder perspective. *The International Review of Retail, Distribution and Consumer Research, 10*(3), 305–318. https://doi.org/10.1080/095939600405992

Wondirad, A., Tolkach, D., & King, B. (2020). Stakeholder collaboration as a major factor for sustainable ecotourism development in developing countries. *Tourism Management, 78*, 104024. https://doi.org/10.1016/j.tourman.2019.104024

World Travel & Tourism Council. (2021). *Global economic impact & trends 2021.*

8

The Role of Smart and Experimental Tourism in Sustainable Development

Fátima L. Carvalho and Silvia Fernandes

Introduction

Research on tourism experiences, big data, social media, interactivity, and innovation has recently expanded to include sustainable tourism and highlights that a new approach emerged between 2018 and 2022. Evidence based on the evolution of related literature shows that virtual tourism and virtual marketing tools for promoting destinations and activities have played a more influential role (Fig. 8.1). Thematic shifts also point to areas such as cultural heritage which, as of 2021, have been linked to smart tourism platforms. Still, little is known about the characteristics of different segments of tourists experiencing cultural tourism

F. L. Carvalho
Research Centre for Tourism, Sustainability and Well-being,
University of Algarve, Faro, Portugal
e-mail: flcarvalho@ualg.pt

S. Fernandes (✉)
Faculty of Economics and CinTurs, University of Algarve, Faro, Portugal
e-mail: sfernan@ualg.pt

T. Walker et al. (eds.), *Sustainable Tourism*,
https://doi.org/10.1007/978-3-031-43528-7_8

Fig. 8.1 Thematic evolution in the literature on smart tourism (Source: Biblioshiny (own elaboration))

attractions through virtual reality applications (Errichiello et al., 2019). There is a need for more research frameworks which seek to understand how best to construct informative and interactive tourist destinations in virtual worlds to attract online and real-world tourists (Huang et al., 2016).

This chapter adopts a bibliometric analysis, using the Bibliometrix R software, which focuses on a selection of 394 documents (concerning smart tourism, experimental research, destination management, virtual reality, contactless technologies, and digital transformation) with a focus on sustainability in tourism (see Table 8.1). It acknowledges that, from 2021, new research experiments have greatly contributed to construct smart tourism systems based on wireless and sensor networks, integrating heterogeneous tourism information through big data algorithms.

Other initiatives aim at refining a travel recommendation system for automating a word-of-mouth (WOM) effect in providing personalized travel-planning services to tourists. Novel tourism tools support the visualization of big data and help to accelerate intelligent tourism awareness of neighboring countries. These also promote cultural exchange through devices such as smart scenic location-based service systems. Thus, emergent debates on sustainable tourism involve two dimensions: one

Table 8.1 Biblioshiny dataset in smart tourism

Description	Results
Timespan	2018–2022
Sources (journals, books, etc.)	191
Documents	394
Average years from publication	1.97
Average citations per documents	6.881
Average citations per year per doc	2.251
References	19554
Document types	
Article	342
Article; early access	48
Article; proceedings paper	4
Document contents	
Keywords Plus (ID)	818
Author's keywords (DE)	1525
Authors	
Number	1155
Author appearances	1286
Authors of single-authored documents	44
Authors of multi-authored documents	1111
Authors collaboration	
Single-authored documents	46
Documents per author	0.341
Authors per document	2.93
Co-authors per documents	3.26
Collaboration index	3.19

dominated by concepts such as experience, co-creation, satisfaction, interactivity, and engagement and another dimension structured around concepts such as big data, innovation, virtual reality, and communication technologies.

This chapter aims to clarify these dimensions and derive lessons towards delineating more assertive strategies in the short and medium term.

A Systematic Literature Review

A systematic literature review identifies, selects, and critically appraises research in order to answer a clearly formulated question (MacKenzie et al., 2012). Thus, regarding the proposed research question, we adopted a bibliometric analysis using the Bibliometrix R software. This allowed us to collect a set of 394 documents concerned with the following keywords: smart tourism, experimental research, destination management, virtual reality, contactless technologies, and digital transformation, with a main link to sustainability in tourism (Table 8.1).

This dataset acknowledges that, from 2021, new research experiments have greatly contributed to construct smart tourism systems based on wireless and sensor networks, integrating heterogeneous tourism information through big data algorithms.

Resulting Thematic Evolution

To consider the resulting thematic evolution, we performed a query in the same software, which produced the output shown in Fig. 8.1.

Figure 8.1 shows that the dataset collected reveals a thematic change to more *experimental* tourism research. This thematic evolution, concerning debates on smart tourism, gives insight into three interesting trends to be studied further: (1) themes formulated as sustainable tourism back in 2018–2020 migrated to new areas of interest such as "virtual tourism" and "smart tourism"; (2) in the post-2021 period, tourism has been studied in relation to "COVID-19" impacts, building a strong association with "virtual tourism"; and (3) in the same period, there is an increased association of "cultural heritage" with "3D modeling" and "smart tourism."

Experimental Research and Innovation

Experimental research/design remains relatively under-utilized in studying tourism-related phenomena (Fong et al., 2017). It has the advantage of using a controlled setting which allows minimizing or ruling out the extraneous effect of other variables and then the ability to enrich theoretical understanding and industry practices. To better characterize this trend, we performed new queries, the outputs of which are

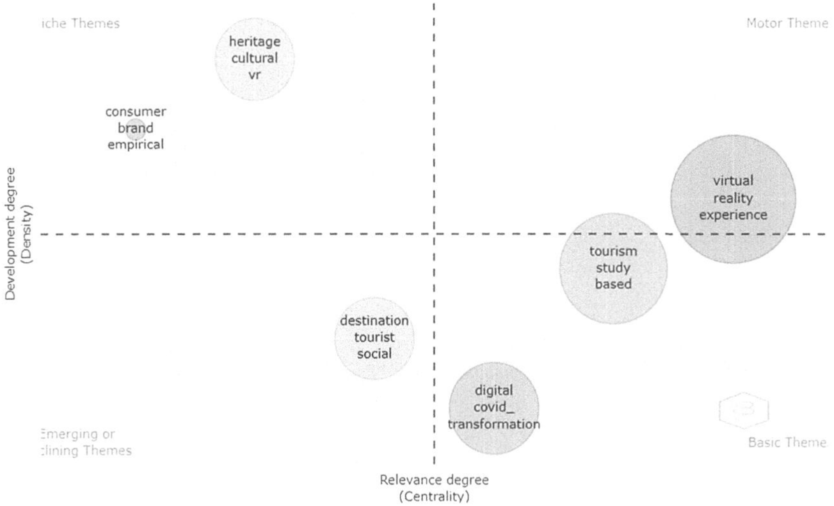

Fig. 8.2 Thematic map of articles on experimental tourism research (Source: Biblioshiny (own elaboration))

presented in Figs. 8.2 and 8.3. Queries refer those options selected during text mining.

Under the same bibliometric analysis, Fig. 8.2 illustrates both a high density and a high relevance of themes such as "virtual reality experience" and, on the other hand, a low concentration of articles on social aspects of tourist destination. Figure 8.3 shows an output from the multiple factorial analysis (keywords plus), with two dimensions/factors. Dimension 1 (horizontal) presents the highest percentage (20.98%) of explanatory power, which has a stronger association between its terms than the other dimension. Terms more related with Dimension 2 are experience, co-creation, interactivity, image, perceived value (cluster 1, red). Dimension 2 (vertical), with less explanatory power (16.58%), is more related with terms such as performance, word-of-mouth, technologies, virtual reality, innovation, social media (cluster 2, blue).

Thus, around the trend to experimental tourism research, practices and principles such as co-creation, cooperation, image, and perceived value have gained more attention.

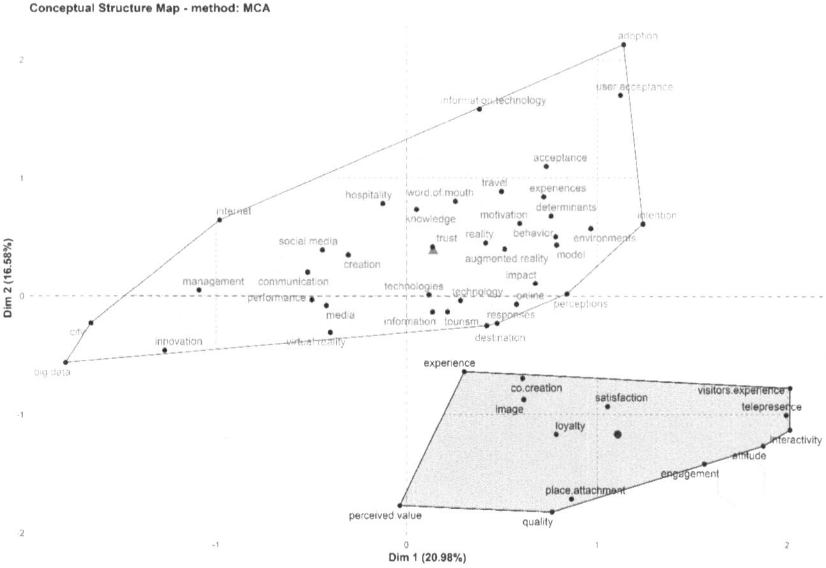

Fig. 8.3 Conceptual structured map on smart tourism (Source: Biblioshiny (own elaboration))

More Interactivity for Higher Value

Tables 8.2 and 8.3 display complementary outputs. Table 8.2 reflects the factorial term map which clarifies the terms' association with Dimensions 1 and 2. It reinforces that although cluster 2 (blue) includes many more terms in the selected literature on smart and experimental tourism, it has less explanatory power, that is, a lower association between its terms (see Fig. 8.3).

Table 8.3 displays the references of the articles collected, showing evidence that a larger part of the literature (cluster 2) focuses on the relations between tourism, information, word-of-mouth, technologies, virtual reality, big data, management, social media, perceptions, and innovation. For instance, the first reference in the table, Tussyadiah and Miller (2019), argues that artificial intelligence (AI) technologies have a great potential to assist, not only in promoting tourism products/services, but also in influencing responsible travel behavior to support sustainability. This

Table 8.2 Word by cluster (Multiple factorial analysis)

Word	Dim.1	Dim.2
Experience	0.3	-0.64
Tourism	0.21	-0.13
Impact	0.67	0.11
Technology	0.28	-0.04
Model	0.79	0.43
Virtual Reality	-0.4	-0.3
Satisfaction	1.05	-0.93
Management	-1.09	0.05
Destination	0.42	-0.25
Travel	0.5	0.89
Augmented reality	0.51	0.4
Information technology	0.38	1.59
Social media	-0.44	0.39
Behavior	0.78	0.5
Intention	1.24	0.61
User acceptance	1.12	1.7
Hospitality	-0.13	0.78
Technologies	0.12	0.01
Information	0.14	-0.13
Internet	-0.98	0.64
Word of mouth	0.25	0.8
Innovation	-1.27	-0.46
Quality	0.76	-1.83
Acceptance	0.73	1.1
Environments	0.97	0.57
Experiences	0.72	0.84
Online	0.58	-0.07
Perceived value	-0.04	-1.77
Reality	0.41	0.45
Co-creation	0.61	-0.69
Knowledge	0.05	0.74
Media	-0.42	-0.08
Performance	-0.5	-0.03
Big data	-1.79	-0.56
City	-1.66	-0.23
Determinants	0.76	0.68
Image	0.61	-0.87
Loyalty	0.78	-1.17
Responses	0.47	-0.23
Telepresence	1.99	-1
Attitude	1.87	-1.27

(continued)

Table 8.2 (continued)

Word	Dim.1	Dim.2
Perceptions	0.84	0.02
Trust	0.14	0.42
Adoption	1.14	2.13
Communication	-0.52	0.21
Creation	-0.31	0.35
Engagement	1.56	-1.42
Interactivity	2.01	-1.13
Motivation	0.59	0.62
Place attachment	0.87	-1.72

Source: Biblioshiny (own elaboration)

Table 8.3 References collected for Biblioshiny analysis

Documents	Dim.1	Dim.2	Contrib	TC
Tussyadiah ip, 2018, tourism manage	0.23	0.2	0.81	204
Hao f, 2020, int j hosp manag	-0.09	-0.03	0.08	120
Kim mj, 2020, j travel res	0.34	0.06	1.03	115
Buhalis d, 2019, j serv manage	-0.14	-0.2	0.52	111
He zy, 2018, tourism manage	0	0.03	0.01	75
Wei w, 2019, tourism manage	0.51	0.47	4.33	71
Bogicevic v, 2019, tourism manage	0.17	-0.06	0.29	66
Loureiro smc, 2020, tourism manage	0.17	-0.01	0.25	64
Marasco a, 2018, j destin mark manage	0.27	0.1	0.72	58
Kim mj, 2019, int j inform manage	0	0.16	0.24	53
Bec a, 2019, tourism manage	-0.12	-0.05	0.16	44
Raptis ge, 2018, int j hum-comput st	-0.12	-0.12	0.27	43
Jin c, 2018, j travel res	-0.04	0	0.01	37
Munoz-leiva f, 2019, physiol behave	-0.17	0.09	0.31	33
Chiao hm, 2018, j hosp leis sport to	0.07	0.46	1.94	31
Schott c, 2018, j comput assist lear	-0.02	0.04	0.02	31
Lee h, 2020, inform manage-amster	0.4	0.12	1.49	30
Li t, 2019, j destin mark manage	0.28	0.13	0.82	30
Ben-shau m, 2018, j travel res	-0.03	0.25	0.57	28
Huang yc, 2019, educ inf technol	0.01	0.05	0.02	28
Errichiello l, 2019, int j tour res	0.28	0.46	2.63	26
Dieck dt, 2018, leisure stud	0.21	0.22	0.83	23
Flavian c, 2019, j travel tour mark	0.01	0.17	0.26	21
Lim h, 2020, j res interact mark	-0.12	-0.11	0.23	21
Leung xy, 2020, int j hosp manag	0.43	-0.13	1.73	20
Preindl r, 2020, supply chain forum	-0.32	0.06	0.93	20
Lee m, 2020, int j hosp manag	0.46	-0.04	1.8	18

(*continued*)

Table 8.3 (continued)

Documents	Dim.1	Dim.2	Contrib	TC
Itani os, 2021, tourism manage	0.16	-0.3	1.07	18
Kim mj, 2020, telemat inform	0.12	0.3	0.95	17
Martin ca, 2018, complexity	-0.13	0.11	0.26	17
Vishwakarma p, 2020, j destin mark manage	-0.06	-0.15	0.24	17
Kizildag m, 2019, int j contemp hosp m	-0.26	-0.07	0.6	16
Tussyadiah i, 2019, ann tourism res	-0.02	0.04	0.01	16
Flavian c, 2021, j hosp market manag	0.47	-0.19	2.24	15
Choi y, 2018, j travel tour mark	0.29	-0.1	0.81	15
Lam c, 2019, int j hosp manag	0.05	0.26	0.63	14
Figueroa v, 2018, int j tour res	-0.16	-0.02	0.23	14
Belli l, 2020, smart cities-basel	-0.54	-0.03	2.55	13
Ogloblina ev, 2020, rev incl	-0.04	-0.2	0.38	13
Jin c, 2018, j clean prod	0.1	0.14	0.26	13
Trunfio m, 2020, curr issues tour	-0.16	-0.02	0.23	13
Hausmann a, 2020, people nat	-0.09	-0.03	0.08	13
Lin lp, 2020, tourism manage	0.15	-0.67	4.35	12
Andriushchenko k, 2020, entrep sustain iss	-0.24	-0.06	0.52	12
Boboc rg, 2019, sustainability-basel	-0.05	-0.11	0.13	12
Barrado-timon da, 2019, sustainability-basel	-0.03	0	0.01	12
Flavian c, 2021, j bus res	0.14	0.07	0.21	11
De bernardi p, 2019, tour anal	-0.21	-0.08	0.45	10
Neidhardt j, 2018, inf technol tour	-0.2	-0.01	0.33	10
Mcleay f, 2019, tourism manage	-0.11	-0.12	0.24	10

study finds three underlying views of AI impacts: Beneficial AI, Destructive AI, and Risky AI. Based on these, three consumer segments are identified: the Laggards, the Enthusiastics, and the Realists. Those belonging to the Realists are likely to respond positively to AI systems, recommending responsible behavior, for a positive role of AI in sustainable tourism.

Having more explanatory power around smart experimental tourism, cluster 1 reveals that principles such as co-creation, cooperation, image, and perceived value have gained more attention. This is in line with more recent works, such as that of Gelter et al. (2022) who stress the need of principles such as stakeholders working together, cooperating, and communicating transparently. These authors defend access to smart platforms and processes, citing trust among municipalities, businesses, public sector, and residents. This also relates to the principles of Dietz et al. (2003) for adaptive management: (1) information be verified, trustworthy, and understood; (2) firms be prepared for change through adaptation, not having too much confidence in current state of knowledge; and (3) maintain face-to-face communication and dense social networks to increase the potential for trust. Thus, making sense of smart tourism development, or in other words understanding new business models change with increased digitization, will improve decision-making processes and communication.

Potential of Smart Systems in Tourism

Today, the internet of things (IoT) has the potential to transform the entire tourism experience. Connecting and selling tourism products and services online is not only price-conscious but also an inspiration for tourists and other stakeholders. Mobile apps, sensors, and location-based tools can give users information and service in seconds. Since more and more tourists expect personalized services, tourism businesses should develop models and tools that can deliver and monitor information that meets customers' needs (Prajapati, 2014).

IoT and AI/AR (artificial intelligence/augmented reality) make it possible to perform activities remotely. These technologies, together with other devices, can connect physical/real objects to the internet. They can bring immense potential to the tourism industry. Table 8.4 shows some examples of resulting features.

These technologies create a smart tourism setting that supplies costumers with relevant information, decision-making support, greater mobility, and enjoyable experiences (Gretzel, 2011; Sigala & Chalkiti, 2014). These smart systems can include a wide range of technologies such as decision-making support systems, context-aware systems, autonomous agents searching for and mining web sources, and augmented realities (Lamsfus et al., 2015). In smart tourism, technology is seen as an infrastructure rather than an individual information system. It is here that experimental research and innovation can propose new approaches and business ideas/models (Lin et al., 2020). This can also provide advanced analytics for more intelligent decisions, as well as options to optimize processes in the entire supply chain involved (Joshi, 2018; Washburn et al., 2010).

Table 8.4 Digital enablers for tourism

Tracking systems	Observing persons/objects on the move and supplying a timely ordered sequence of location data for further processing
Personalization	Suggestions based on tourist preferences, tailored service/product to answer specific needs
Less off the beaten path	Explore more of what one likes
Destination environments	By cameras and video tools and analytics
Real time updates	About points of interest, conditions (clean & safe, other), currency rates, etc.
Augmented reality	Overlaying icons to simulate, experiment, integrate
Targeted information	Broadcast useful information to the smartphone

Source: Adapted from KSTDC (2017)

The main impacts of these cutting-edge technologies are that they (1) provide tools to respond accurately and remotely to the users and (2) enable customers to become co-creators/innovators (Du, 2021). The Social Context Mobile (SoCoMo) conceptual model (Buhalis & Foerste, 2015), for example, is a framework that enables marketers to increase value for all stakeholders through mobile marketing. Customers also now have access to apps and tools to co-create the influence of which is amplified by social media. This is very interesting because customers provide ideas and insights into hidden problems, market opportunities, and innovation at scale. Once this relationship is built, they can be used as early product/service promoters and testers.

This requires a significant shift involving changes to business models and organizational structure and culture (Alvarado-Uribe et al., 2018; Menzies, 2016). When tourism and hospitality firms can offer digital platforms for customers to participate in value creation, they define stronger path-dependent and connecting behavior. For example, tourists' involvement in the process of co-creating a trip increases satisfaction, which leads to more positive remembrance of the whole experience (Shaw et al., 2011). Co-creation is increasingly important to tourism and other areas as it includes or involves customers' ideas and preferences in the conception of products/services offering them what they really want (Buonincontri & Micera, 2016).

Another result of co-creation can be the increased engagement of tourists with residents within community activities at the destination. This change involves shifting from traditional enterprise architecture to digital architecture which addresses features such as networked community capabilities (interacting with stakeholders), globalization (borderless firms), product/service innovation (virtual reality), collaboration (mobile services/work), and flexibility. In smart systems, a wide variety of sensors can be used to obtain and use contextual information. Examples of sensors used are GPS (for location, speed), light and vision (to detect objects, activities), microphones (for data about sound, talking), gyroscopes (for movement, orientation), magnetic sensors (to determine orientation), and temperature sensors (to assess environments) (Schmidt, 2013). Other sensors serve to detect the physiological context of the user, such as galvanic skin response. All types of sensors available can be used to feed these systems with context/real-time information.

Examples and Cases

The Portuguese tourism industry should follow emerging technological trends for greater innovation toward the growth of online services. The widespread use of mobile cyber-physical platforms is an example, whose attributes deal with the critical role that time and place have in today's communications. Some examples are the smart wearable devices that can give new options to their users or other devices that control and execute activities or processes such as on-time delivery. These platforms also provide mechanisms for the selection and alignment of processes that capture more characteristics of a place and reflect its changes. These attributes can change the way of developing products and doing activities.

Another example is a system that can collect new relevant information from tourists' feedback in real time through a mobile app. Then, a data analysis of the dataset collected can quickly perform the action or deliver the information needed. The devices mentioned here can include cameras, sensors, geolocation equipment, and even real objects (such as clothes, objects in a museum, in a room, a wall, etc.); that is why they are called "cyber-physical." For instance, the AirQ cyber-physical system developed by SmartSense in Croatia for continuous air quality monitoring allows tourists to provide real-time feedback through a mobile application. It collects relevant information on air quality and then conducts an analysis of the dataset obtained that can allow the system to perform either preventive or corrective actions.

The presence of sensors in equipment or real objects allows the collection of important data, the collection of which is relevant to ensure that up-to-date information is available to the users. The AirQ sensor can even be positioned in locations previously inaccessible and new sensors can be developed to monitor additional areas. These can continue monitoring the full range of pollutants and sensors deployed to collect additional information such as conditions that affect air quality daily. This system offers enough bandwidth to transmit all collected data as needed. This can lead to a better customer experience, and costs of maintenance and installation are reduced. The use of mobile applications enhances tourist relationships and, in turn, social capital through reciprocity and cooperation (Barile et al., 2014; Kim & Kim, 2017).

New use cases and business models can be conceived based on the data the AirQ system obtains (Belli et al., 2020). This data will not only be used to generate direct revenue for exceeding air quality levels, but also indirectly when automatic actions are triggered. In turn, traffic management systems can then be controlled to ease pollution for a better user experience. Smart solutions can therefore lead to greater changes in terms of pollution management and mitigation. This presents a valuable opportunity for cities wishing to be smart and sustainable.

Based on this kind of process, several other projects can be developed, with impacts on the tourism sector. For example, mobile operators are now well-informed in developing and providing tools for better crowd management. Tracking the location of users' mobile phones and analyzing data collected by mobile-enabled sensors can provide an accurate way to manage and monitor all sorts of gatherings, whether they are in city centers or in remote locations (GSMA, 2018). These and other potential applications become highly relevant today facing new climate and social challenges (environmental threats, epidemics, terrorism, etc.).

Other projects relate to recommendation systems which enable users to find what they need within large data collections and supporting a variety of applications of e-tourism (Su et al., 2019). A tourism recommendation system is a crucial solution to help tourists discover more diverse tourism destinations; a content-based approach in such a system can be an effective way of recommending items because it looks at the user's preference histories (Fudholi et al., 2021; Kontogianni et al., 2022). This is experimental tourism research at its highest level as disruptive technologies which, like IoT and augmented/virtual reality, may match the need for future contactless services in unique situations (Huang & Liu, 2021).

Portuguese Digital Context

In acknowledging how Portugal performs in matters of smart/experimental tourism, the following figures illustrate the main performance-related indicators. Figure 8.4 shows, in each country considered, the percentage of enterprises that offered customized or co-created goods/services. Portugal has registered, in these variables, lower values than several other countries.

In relation to channels that acquire relevant information for innovation (Fig. 8.5), Portugal shows lower levels especially in published patents, reverse engineering, and crowdsourcing. The variables that were dominant until recently are mainly related to open-innovation (OI) practices. OI is very important to cope with the current challenges of smart development and supporting technologies. It involves having firms' boundaries open to internal/external paths of sharing one's own or anothers' ideas, risks, innovations, and IP (intellectual property). Thus, Portugal is still catching up in this regard through increased initiatives surrounding regional/national innovation accelerator programs (Fernandes et al., 2017).

This lag is in line with the country's reality in terms of its national innovative performance. An example is its national innovation index (Sii—Summary innovation index). The report that presents this indicator is the European Innovation Scoreboard (EIS), which has been

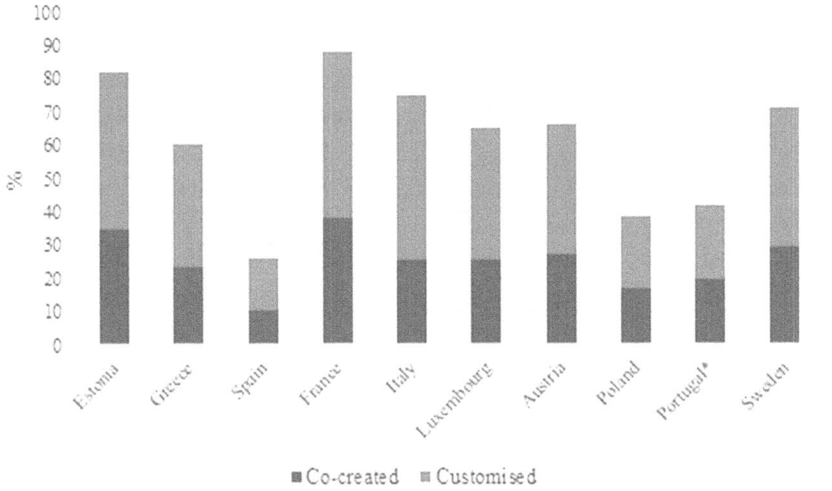

Fig. 8.4 Enterprises that offered customized or co-created goods/services (Source: own elaboration, based on Eurostat (2018))

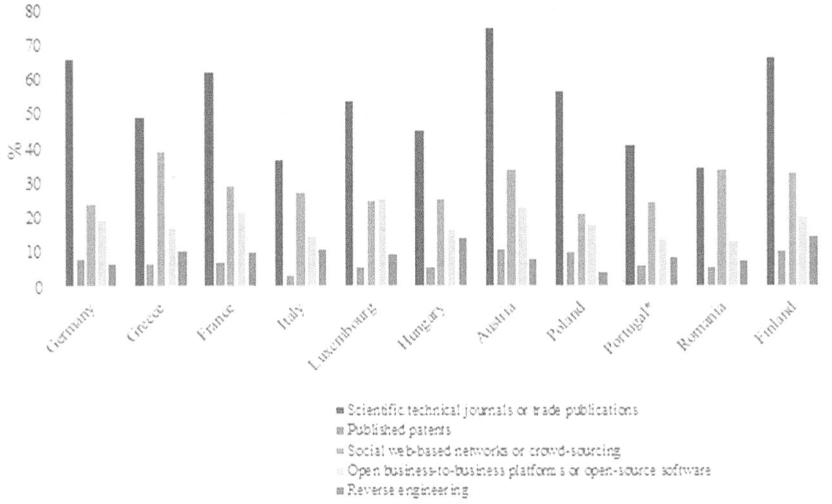

Fig. 8.5 Enterprises by kind of channels to acquire relevant information for innovation (Source: own elaboration, based on Eurostat (2018))

published annually since 2001. It provides a comparative analysis of innovation performance in the member states of the European Union (EU) assessing the strengths and weaknesses of their national innovation systems. This has helped member states identify areas that need further intervention. According to this index, the EIS assigns four possible classifications: Leading Innovators, Strong Innovators, Moderate Innovators, and Modest Innovators.

In recent years, Portugal has been classified as a Moderate Innovator. The first year it was considered a Strong Innovator was 2020 (in 12th place). That year, Portugal presented values above the European average in variables such as the creation of new companies and entrepreneurship activities. Regarding the basic conditions for innovation, particularly in terms of attractiveness of the research system and broadband penetration, Portugal is above average. However, variables that reflect its ability to transfer knowledge, such as patents and the export of innovative products, are still below the average, although positive developments are evidenced.

Thus, Portugal has recently moved from being a "strong" to a "moderate" innovator (EIS, 2021). This assessment is based on the following metrics (Table 8.5), which make up the referred *Sii* index:

In this composite indicator, the strongest dimensions (or variables) in the Portuguese innovation ecosystem are:

Table 8.5 Dimensions of the Sii in Portugal

Innovation in Portugal 2021	Value
Sii (Summary innovation index)	80.2
Human resources	94.6
Attractive research systems	115.6
Digitalization	121.9
Finance and support	94.1
Firm investments	50.1
Use of information technologies	103.5
Innovators	79.2
Linkages	102.4
Intellectual assets	76.7
Employment impacts	65.2
Sales impacts	64.8
Environmental sustainability	38.9

Source: EIS (European Innovation Scoreboard) (2021)

- Favorable innovation environment
- Attractive search systems
- Broadband expansion
- Number of foreign doctoral students

Indeed, the country is doing well in terms of digital intensity (e.g., broadband expansion) as Fig. 8.6 also shows.

Nevertheless, other important dimensions must be analyzed and improved, as their performance is weak in the country's national innovation ecosystem, such as:

- Expansion of connections
- Intellectual assets
- Exports of knowledge-intensive services
- Innovation and internal technology adoption
- Private co-financing of public expenditure on R&D

Portugal offers a favorable environment for innovation and talent attraction. However, it is relatively behind in the conversion of acquired knowledge into sales and exports. One motive for this can be the level of

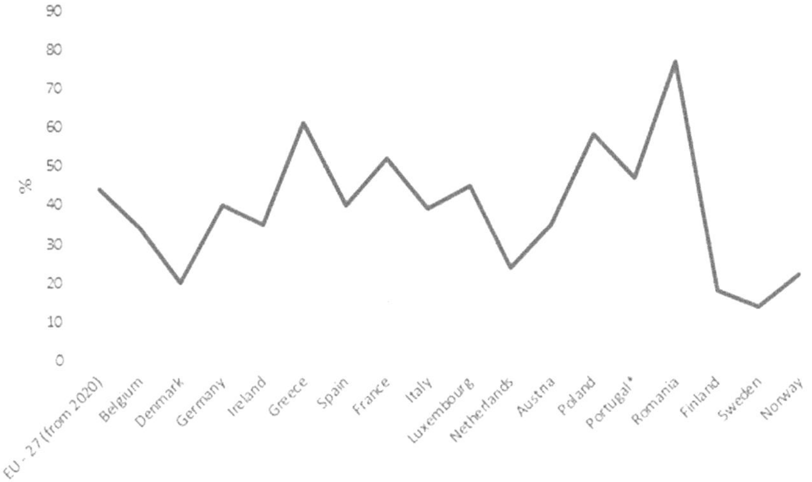

Fig. 8.6 Digital intensity (Source: own elaboration, based on Eurostat (2020))

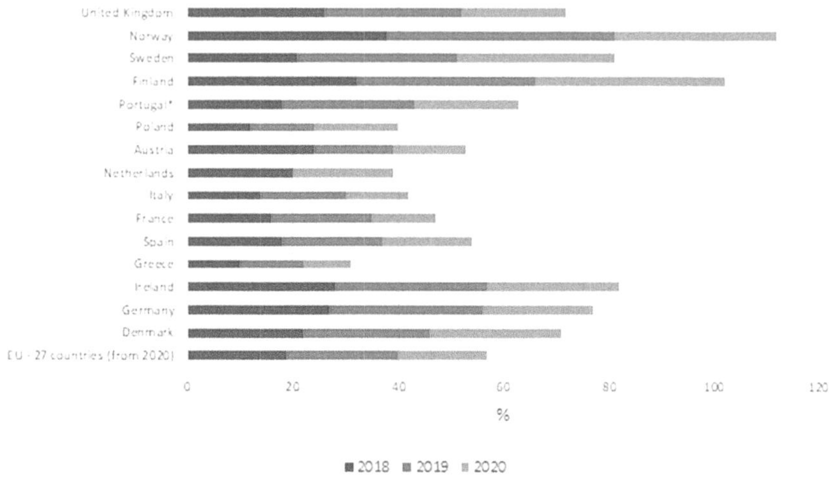

Fig. 8.7 Enterprises that provided training to their employees to develop their ICT skills

specialized training in ICT (Information and Communication Technologies) provided by enterprises to their employees (Fig. 8.7).

This reality, including the pandemic impact, may delay tourism recovery due to cultural and digital challenges ahead.

The role of universities and related innovation accelerators, which are stimulating relations with enterprises (especially small businesses), is increasingly relevant as they expand their links with companies to create innovative products/services and generate employment (Allahar & Sookram, 2019). The emphasis is placed on the need to develop strong collaboration to build a sustainable quadruple ecosystem (university-business-government-society). We can observe an example of this in the south of Portugal (Algarve), which is rapidly becoming a creative hub for Portugal (Algarve Tech Hub, 2021) thanks to its balance between quality of life and work, as well as between its university and companies, entrepreneurs, investors, and start-ups alike are easily starting and "taking a company off the paper." This growth is furthered by the influential role of its recent Technological Hub (innovation accelerator), which gives access to firms seeking local and foreign infrastructure and talent.

Conclusion

Recent literature and trends suggest that sustainability may become more prominent in tourism choices, due to a greater awareness of the adverse impacts of conventional tourism. Natural areas and smart destinations are expected to drive tourism recovery and may result in lower environmental impacts. The literature and metrics review of this chapter reveal a pace of change that tourism researchers have recently acknowledged and analyzed concerning this important sector of the economy. Existing knowledge might be subject to a paradigm shift due to disruptions from the COVID-19 pandemic (Zenker & Kock, 2020), which has changed attitudes and behaviors, and tourism models and activities need to be dynamic and adaptive to cope with this and other future disruptors. Therefore, more studies are needed which include behavioral theory and emergent technologies assessment. This is important both for analytical reasons and for the enhancement of smart and contactless products/services. The increasingly dynamic and complex challenges in the field of sustainable tourism call for more cooperative research on the factors that influence smart tourism development.

Portugal, according to its Sii (Summary innovation index) in 2019 was considered as a "moderate" innovator and became a "strong" innovator in 2020, before its recent return to a "moderate" classification (EIS, 2021). The country's main innovation strengths are attractive research systems and a strong use of information technologies (IT). The recent decline of its innovation performance is mainly due to below-average levels in supporting internal innovation and IT specialists. In other words, there is a weakness in the use/application of the results from its Research and Development. This profile will influence the type, direction, and scale of firms' innovation, including in the tourism sector.

The approach we use in this chapter has a limited scope and future work should deepen their analysis of trust and safety issues and explore how traveler behavior evolves in a post-COVID context. This requires studies with innovative mixed methods, combining systems theory with new methods for data collection and analysis. Future studies should consider the ways in which consumer trends will be reshaped in the face of the pandemic, which new market segments or niches will emerge, as well as how smart technologies and systems can deliver these new emerging services.

Studies with a greater focus on safe and contactless tourism experiences will also make a relevant contribution. These should address whether safety and trust will play a major role in destination decisions, which policies can ensure investments in the tourism industry, address quality and sustainability, as well as how tourism could experiment with new uses of digitization.

It is important to analyze how businesses can translate crisis into transformative innovation. Tourism research has never felt the need to hold its purpose as much as today.

Acknowledgments This paper is financed by National Funds provided by FCT-Foundation for Science and Technology through project UIDB/04020/2020.

References

Algarve Tech Hub. (2021). *The ecosystem.* https://algarvetechhub.com/about/the-ecosystem/

Allahar, H., & Sookram, R. (2019). A university business school as an entrepreneurial ecosystem hub. *Technology Innovation Management Review, 9*(11), 15–25. https://doi.org/10.22215/timreview/1280

Alvarado-Uribe, J., Gómez-Oliva, A., Molina, G., Gonzalez-Mendoza, M., Parra-Meroño, M., & Jara, A. (2018). Towards the development of a smart tourism application based on smart POI and recommendation algorithms: Ceutí as a study case. In L. Barolli & T. Enokido (Eds.), *Innovative mobile and internet services in ubiquitous computing. Advances in intelligent systems and computing*. Springer. https://doi.org/10.1007/978-3-319-61542-4_92

Barile, F., Calandra, D., Caso, A., DAuria, D., Mauro, D., Cutugno, F., & Rossi, S. (2014). ICT solutions for the ORCHESTRA project: From personalized selection to enhanced fruition of cultural heritage data. In *10th international conference on signal-image technology and internet-based systems* (pp. 501–507). https://doi.org/10.1109/SITIS.2014.12

Belli, L., Cilfone, A., Davoli, L., Ferrari, G., Adorni, P., Di Nocera, F., & Dall'Olio, A., Pellegrini, C., Mordacci, M., & Bertolotti, E. (2020). IoT-enabled smart sustainable cities: Challenges and approaches. *Smart Cities, 3*(3), 1039–1071. https://doi.org/10.3390/smartcities3030052

Buhalis, D., & Foerste, M. (2015). SoCoMo marketing for travel and tourism: Empowering co-creation of value. *Journal of Destination Marketing & Management, 4*(3), 151–161. https://doi.org/10.1016/j.jdmm.2015.04.001

Buonincontri, P., & Micera, R. (2016). The experience co-creation in smart tourism destinations: A multiple case analysis of European destinations. *Information Technology & Tourism, 16*, 285–315. https://doi.org/10.1007/s40558-016-0060-5

Dietz, T., Ostrom, E., & Stern, P. C. (2003). The struggle to govern the commons. *Science, 302*, 1907–1912. https://doi.org/10.1126/science.1091015

Du, J. (2021). Research on intelligent tourism information system based on data mining algorithm. *Mobile Information Systems, 2021*. https://doi.org/10.1155/2021/5727788

EIS, European Innovation Scoreboard. (2021). *Internal market, industry, entrepreneurship and SMEs*. https://ec.europa.eu/growth/industry/policy/innovation/scoreboards_en

Errichiello, L., Micera, R., Atzeni, M., & Del Chiappa, G. (2019). Exploring the implications of wearable virtual reality technology for museum visitors' experience: A cluster analysis. *International Journal of Tourism Research, 21*, 590–605. https://doi.org/10.1002/jtr.2283

Eurostat. (2021). *Digital economy and society—Benchmarking digital Europe: key performance indicators*. https://ec.europa.eu/eurostat/data/database

Fernandes, S., Cesário, M., & Barata, J. M. (2017). Ways to open innovation: Main agents and sources in the Portuguese case. *Technology in Society, 51*, 153–162. https://doi.org/10.1016/j.techsoc.2017.09.002

Fong, S., Lo, M., Songan, P., & Nair, V. (2017). Self-efficacy and sustainable rural tourism development: Local communities' perspectives from Kuching, Sarawak. *Asia Pacific Journal of Tourism Research, 22*(2), 147–159. https://doi.org/10.1080/10941665.2016.1208668

Fudholi, D., Rani, S., Arifin, D., & Satyatama, M. (2021). Deep learning-based mobile tourism recommender system. *Scientific Journal of Informatics, 8*(1), 111–118. https://doi.org/10.15294/sji.v8i1.29262

Gelter, J., Fuchs, M., & Lexhagen, M. (2022). Making sense of smart tourism destinations: A qualitative text analysis from Sweden. *Journal of Destination Marketing & Management, 23*. https://doi.org/10.1016/j.jdmm.2022.100690

Gretzel, U. (2011). Intelligent systems in tourism: A social science perspective. *Annals of Tourism Research, 38*(3), 757–779. https://doi.org/10.1016/j.annals.2011.04.014

GSMA. (2018). *NB-IoT: Internet of Things case study series—Smart cities*. https://www.gsma.com/iot/wp-content/uploads/2018/02/iot_dt_airq_01_18.pdf

Huang, T., & Liu, B. (2021). Augmented reality is human-like: How the humanizing experience inspires destination brand love. *Technological Forecasting and Social Change, 170*. https://doi.org/10.1016/j.techfore.2021.120853

Huang, Y., Backman, K., Backman, S., & Chang, L. (2016). Exploring the implications of virtual reality technology in tourism marketing: An integrated research framework. *International Journal of Tourism Research, 18*, 116–128. https://doi.org/10.1002/jtr.2038

Joshi, S. (2018). Social network analysis in smart tourism driven service distribution channels: Evidence from tourism supply chain of Uttarakhand, India. *International Journal of Digital Culture and Electronic Tourism, 2*(4), 255–272. https://doi.org/10.1504/IJDCET.2018.092199

Kim, D., & Kim, S. (2017). The role of mobile technology in tourism: Patents, articles, news, and mobile tour app reviews. *Sustainability, 9*, 1–45. https://doi.org/10.3390/su9112082

Kontogianni, A., Alepis, E., & Patsakis, C. (2022). Promoting smart tourism personalized services via a combination of deep learning techniques. *Expert Systems with Applications, 187*. https://doi.org/10.1016/j.eswa.2021.115964

KSTDC. (2017). *India Tourism Statistics*. https://tourism.gov.in/sites/default/files/2020-04/INDIA%20TOURISM%20STATISTICS%202017.pdf

Lamsfus, C., Wang, D., Alzua-Sorzabal, A., & Xiang, Z. (2015). Going mobile defining context for on-the-go travelers. *Journal of Travel Research, 54*(6), 691–701.

Lin, M., Li, F., & Zhou, H. (2020). A research on the combination of oblique photography and mobile applications based on the sustainable development of tourism. *Sustainability, 12*. https://doi.org/10.3390/su12093501

MacKenzie, H., Dewey, A., Drahota, A., Kilburn, S., Kalra, P., Fogg, C., & Zachariah, D. (2012). Systematic reviews: What they are, why they are important, and how to get involved. *Journal of Clinical and Preventive Cardiology, 1*(4), 193–202.

Menzies, A. (2016). *Co-creation for successful customer engagement*. http://www.summa.com/blog/successful-customer-engagement-with-co-creation

Prajapati, V. (2014). Mobile technology advancements speed up the growth of tourism worldwide. http://www.techprevue.com/mobile-technology-advancementsspeedup-the-growth-of-tourism/

Schmidt. (2013). *Context-aware computing*. https://www.interaction-design.org/literature/book/the-encyclopedia-of-human-computer-interaction-2nd-ed/context-aware-computing-context-awareness-context-aware-user-interfaces-and-implicit-interaction

Shaw, G., Bailey, A., & Williams, A. (2011). Aspects of service-dominant logic and its implications for tourism management: Examples from the hotel industry. *Tourism Management, 32*(2), 207–214. https://doi.org/10.1016/j.tourman.2010.05.020

Sigala, M., & Chalkiti, K. (2014). Investigating the exploitation of web 2.0 for knowledge management in the Greek tourism industry: A utilization-importance analysis. *Computers in Human Behavior, 30*, 800–812. https://doi.org/10.1016/j.chb.2013.05.032

Su, X., Sperlì, G., Moscato, V., Picariello, A., Esposito, C., & Choi, C. (2019). An edge intelligence empowered recommender system enabling cultural heritage applications. *IEEE Transactions on Industrial Informatics, 15*(7), 4266–4275. https://doi.org/10.1109/TII.2019.2908056

Tussyadiah, I., & Miller, G. (2019). Perceived impacts of artificial intelligence and responses to positive behaviour change intervention. In J. Pesonen & J. Neidhardt (Eds.), *Information and communication Technologies in Tourism* (pp. 359–370). Springer. https://doi.org/10.1007/978-3-030-05940-8_28

Washburn, D., Sindhu, U., Balaouras, S., Dines, R. A., Hayes, N., & Nelson, L. (2010). *Helping CIOs understand 'smart city' initiatives*. http://www.uwforum.org/upload/board/forrester_help_cios_smart_city.pdf

Zenker, S., & Kock, F. (2020). The coronavirus pandemic—A critical discussion of a tourism research agenda. *Tourism Management, 81*, 104164. https://doi.org/10.1016/j.tourman.2020.104164

9

A New Tool for Smart Tourism Destinations: Indicators of Sustainable Intelligence

Yaiza López-Sánchez, Juan Ignacio Pulido-Fernández, and José Luis Durán-Román

Introduction

The emergence of Smart Tourism Destinations/Smart Destinations (SDs), a new approach to the planning and management of tourist destinations, has now adopted sustainability as a crucial part of its framework. Sustainability Indicators are a fundamental tool for destinations developing and implementing sustainable tourism principles and policies in real-world contexts. This chapter contributes to a new understanding of sustainability and tourism, specifically with respect to demand analysis. In order to better integrate tourists in the process of driving sustainable

Y. López-Sánchez (✉)
Laboratory of Analysis and Innovation in Tourism (LAInnTUR),
University of Jaén, Jaén, Spain

Department of Applied Economics (Political economy),
University of Málaga, Málaga, Spain
e-mail: ylopez@uma.es

T. Walker et al. (eds.), *Sustainable Tourism*,
https://doi.org/10.1007/978-3-031-43528-7_9

SDs, it is first necessary to evaluate and measure the tourists' attitudes, knowledge, behaviors and level of commitment regarding sustainability. We propose to do this via a novel concept called sustainable intelligence, an element found within the framework of smart sustainability.

The relationship between SDs and Sustainable Tourism Indicators has been extensively studied, but there is a lack of information regarding the incorporation of tourists into such tools. Considering this gap in the literature, we propose a new tool: Indicators of Tourists' Sustainable Intelligence, as a way to integrate tourists into the advancements made by SDs toward sustainability. Tourist behavior is a fundamental issue in tourism studies and, in particular, pro-sustainable tourist behavior is thriving as a field of study. Reviewing the literature, many definitions have developed around this idea, both when it comes to segmenting tourists and in the denomination of different typologies. The scientific debate regarding consumer behavior and sustainability issues is heated and, in many ways, controversial. It is productive to examine the relationships between the values and attitudes of tourists and their actual behaviors with regard to sustainable consumption patterns. One significant problem lies in the number and kind of variables to be considered. The variables selected by researchers differ according to the researcher's understanding of sustainability. This chapter proposes that tourists are not a homogenous and static group but instead vary, depending on their level of sustainable intelligence. This idea includes several variables related to the commitment, attitude, knowledge, and behavior of tourists regarding the suitability of a destination.

J. I. Pulido-Fernández
Laboratory of Analysis and Innovation in Tourism (LAInnTUR), University of Jaén, Jaén, Spain

Department of Applied Economics, University of Jaén, Jaén, Spain
e-mail: jipulido@ujaen.es

J. L. Durán-Román
Department of Business Organization, Marketing and Sociology, University of Jaén, Jaén, Spain
e-mail: jduran@ujaen.es

Recent literature defines the sustainable intelligence of tourists as their ability to apply personal experience and knowledge regarding the effects of tourism on their environment, developing proactive behavior toward sustainable tourism, from the perspective of both consumption and production. From a statistical perspective, sustainable intelligence is a latent variable (not directly observable); therefore, it has to be quantified by indirect indicators, which, to a greater or lesser degree, are related to the latent variable (López-Sánchez & Pulido-Fernández, 2016). This chapter presents the results of a three-round Delphi survey conducted into expert opinion on the development of Indicators of Tourists' Sustainable Intelligence in order to lay the necessary foundations to define and measure the different dimensions of tourists' *sustainable intelligence* and the development of a global indicator, related to the commitment, attitude, knowledge, and behavior of tourists related to sustainability. This tool incorporates the demand perspective for the sustainable management of SDs as a key factor in the move toward sustainability.

Smart Destinations or Smart Sustainable Destinations?

Technological advances and their impact on society, changes in consumption patterns, and new business models—in short, the new era of information and communication technologies—mean that tourist destinations have to face a set of new challenges arising from changes in both demand (tourists) and supply (destinations), influenced by new technologies.

The terms "smart cities" and "smart destinations" have become increasingly popular in the last decade, and interest in these concepts and their real-world applications continues to grow. Based on an extensive review of the literature, the origins of the term "smart destinations" emerges from the idea of "smart cities" (Buhalis & Amaranggana, 2014). The smart city concept covers a variety of industries, including tourism (Guo et al., 2014) and smart cities support and enable the establishment of smart tourist destinations (Jasrotia & Gangotia, 2018). The smart city concept itself is still in flux and subject to debate, varying according to geopolitical context. On one hand, it has been defined as the new urban utopia (Datta, 2015) and, on the other, as the biggest driver for economic growth since the industrial revolution (Gupta & Hall, 2017). The concept is still evolving and has not yet achieved mainstream recognition across the globe due to economical, technological, and governance restraints (Silva et al., 2018). Most studies describe smart cities as multi-dimensional systems (Walter et al., 2015) that revolve around digital innovation to improve the well-being of citizens and the efficiency of urban services, generating new economic opportunities in cities.

According to the OECD (2020), smart cities are defined as initiatives or approaches that effectively leverage digitalization to boost citizen well-being and deliver more efficient, sustainable, and inclusive urban services and environments as an element of a collaborative multi-stakeholder process. Depending on the authors, however, each city is smart as long as it is committed to the implementation of a smart economy (competitiveness), smart environment (natural resources preservation), smart governance (participation), smart living (quality of life), smart mobility (transport and ICT), and smart people (social and human capital) (Benevolo et al., 2016). An additional aspect worth considering is that without sustainability, a city (or destination) cannot be conceived of as smart (Perles-Ribes & Ivars-Baidal, 2018). The connections between smartness and sustainability are expressed on two complementary levels: the strategy of the destination and the application of technologies to create more efficient environmental management (Ivars-Baidal et al., 2019). The sustainability of smart destination models should not be limited merely to applying technologies to improve traditional processes but should also include management for sustainability (encompassing all of its pillars) or, even more radically, how to evolve toward steady state or degrowth situations (Fletcher et al., 2019).

Reviewing the concepts of sustainability and smartness as applied to tourist destinations, it is possible to identify common and complementary aspects that allow for higher levels of sustainability based on the development of a smart destination strategy (based on Perles-Ribes & Ramón-Rodríguez, 2017). Several authors have analyzed the role of sustainability in smart city strategies from different perspectives and concluded that it is under-utilized and should be better integrated into smart city projects (Bibri & Krogstie, 2017; Haarstad, 2017). Some authors have even recommended the use of "smart sustainable cities" as a more accurate term in order to emphasize the sustainability dimension. The assessment of smart city performance should not only use output indicators that measure the efficient deployment of smart solutions, but also impact indicators that measure contribution toward the final goals of environmental, economic, and social sustainability (Ahvenniemi et al., 2017).

In short, a linear relationship between smartness and sustainability cannot be conceived. Perles-Ribes and Ivars-Baidal (2018) analyzed the sustainability and smartness of tourist destinations. In this study, they propose the concept of smart sustainability, "oriented more towards paradigms of strong sustainability which impose limits to growth and implement the principles of the circular economy as opposed to the linear growth approaches, which are becoming completely obsolete given the growing pressure on natural resources and the threat of climate change" (p. 166). The field of knowledge that studies the integration and dependencies of sustainability and destination intelligence still has many unknowns to be solved and many solutions to be offered. There are several areas which merit further exploration by those who research the design of smart destinations (and smart tourism more generally). One key issue is the need for further research on how tourists, in general, respond to the opportunities and uncertainties associated with smart innovations or smart sustainability (Williams et al., 2020).

The Role of Tourists in Smart Sustainable Destinations: Tourists' Sustainable Intelligence

Smart destinations and smart sustainability are gaining attention as new frameworks with which to understand the advances in information and communication technologies and their impact on the relationship between destinations and tourists. There is, however, a fundamental element of the smart tourism ecosystem that has been neglected in the research thus far: the tourists themselves (Femenia-Serra et al., 2018). There is scientific consensus that citizen engagement is a key element in most definitions of smart cities (Granier & Kudo, 2016). Given that the smart destination derives from the smart city, tourists should occupy a central position in its examination. One of the most important areas of focus of the smart city is its citizens (regular, permanent residents), but these approaches do not always include short-term tourists in the design and creation of new communication channels. Just as it is a major challenge to understand the behavior of citizens of smart cities (Chourabi et al., 2012), it is also essential to understand the behavior of tourists in a smart and sustainable destinations as well. Moreover, as this study proposes, we must consider not only their behavior, but the knowledge, attitude, consumption, and economic commitment of tourists with respect to the sustainability of destinations merits consideration. This begs the question: what defines sustainable intelligence for tourists, and can it be measured?

The term sustainable intelligence, applied in this case to tourists, refers to the capacity of tourists to apply their experience and knowledge in order to mitigate the negative effects of their activities, and to develop a proactive stance toward sustainable tourism. A tourist with a high level of sustainable intelligence is one who demonstrates empathy, action, and commitment toward sustainable tourism development at the destination where they enjoy their holidays. These kinds of tourists will have a real understanding of the meaning and practical implications of sustainability. Progress toward understanding and measuring the sustainable intelligence of tourists is fundamental to understanding the motivations, expectations, and behaviors of the tourists of today, to be able to attract and maintain a demand for tourism that contributes to real progress in the sustainability of destinations (López-Sánchez & Pulido-Fernández, 2016).

Methodological Framework

The empirical part of this study consists of a Delphi analysis with the aim of ascertaining the opinion of a selected group of experts as to what should be the main dimensions and global or generic indicators when assessing and measuring the sustainable intelligence of tourists. The current lack of consensus on how to measure the sustainable intelligence of tourists prevents valid and reliable measurements. Our research lays the groundwork for the creation of a standardized instrument and the development of a battery of measurable and objective indicators. Our study provides a starting point for developing a set of indicators to measure the sustainable intelligence of tourists. In order to accomplish this task, it is first essential to reach a consensus on the main dimensions that define the concept and, next, to determine the global indicators that give content to each dimension. Once the foundations of this methodological tool have been laid, future research will be required in order to (1) identify the specific indicators (which will give specific content to each global indicator) and (2) determine the relative weight of each specific and global indicator, as well as of each dimension, in the result.

The Delphi research method (whose name derives from the Oracle of Delphi in Greece) is based on the principle of collective intelligence and aims to seek consensus from the opinions of a group of experts on the subject to be investigated. This is a prospective technique for obtaining qualitative or subjective information. It results from combining the knowledge and experience of experts in various fields and then statistically quantifying these opinions using the mean, median, and quartiles. The Delphi research method was specifically designed as a forecasting tool for the Rand Corporation in the 1950s. However, over the last several decades, Delphi research has been more frequently used to facilitate group communication for decision-making and planning. Its origin is based on the idea that two heads are better than one (Dalkey et al., 1972), that is, that the collective judgement of a group of experts is better than that of a single expert (Kaynak & Macaulay, 1984). Linstone and Turoff (2002) defined it as a method for structuring a group communication process whose effectiveness is determined by allowing a group of individuals to be treated as a whole in order to solve a complex problem. Its aim is to take advantage of "group" methods and to reduce the negative (mainly psychological) interaction effects of group or face-to-face meetings.

We chose the Delphi method because it relies on the need for information sharing and reasoning among the members of the expert group, which ensures a diversity of judgements and opinions that can be effectively and practically incorporated into the research process at a reduced cost (Pill, 1971). A large body of research recognizes the value of Delphi analysis as an effective process of expert opinion and knowledge collection (Adler & Ziglio, 1996; Green et al., 1990). The main characteristics of this method are summarized in Donohoe and Needham, 2009 and Landeta, 2006:

(a) Participant anonymity: The method maintains the anonymity of the participants as participant responses go directly to the group coordinator. The method also allows for a group working process with experts who are not present in the same time or space. It also aims to avoid the negative influence of factors in the individual answers such as the personality and status of the participating experts.
(b) Controlled feedback: The exchange of information between the experts is mediated through the use of a study group coordinator, so that all irrelevant information can be eliminated.
(c) Statistical group response: All opinions form a part of the final answer. Questions are formulated in a manner that allows for the answers to be processed quantitatively and statistically.

The successful implementation of this research method depends on the panel members and the panel leaders; moreover, the process is most effective when the topics to be investigated lead to subjective judgements rather than precise analytical techniques. The process consists of collecting the opinions of experts anonymously by means of a questionnaire, analyzing the answers, and returning the revised questionnaires back to the authors so that they can modify their opinions if it is deemed appropriate (controlled feedback). The procedure is repeated in successive rounds (in this study there were three rounds) until the answers are refined and a consensus is reached. Research has demonstrated that the Delphi method is a suitable method for application in tourism research (Donohoe & Needham, 2009; Green et al., 1990; Lin & Song, 2015; Pulido-Fernández, 2005). The number of experts participating in a Delphi panel usually ranges from 10 to 30 (Landeta, 2006). A high number of participants increases reliability, but, if the experts are very homogeneous, it does not make sense to increase the number, because few new ideas are contributed, and the increased volume makes it more difficult to analyze the results.

In Delphi-based surveys, the minimum number of participants is fewer. Participants are not selected at random, but because of the need for particular expertize. There have been numerous successful Delphi studies conducted with the initial participation of only ten experts (Dalkey et al., 1972; Dalkey & Helmer, 1963). In the Delphi method, the recursion produced by three or more rounds, the degree of expertize, the anonymity of participants, and the opportunity for experts to re-evaluate and revise answers based on having seen the opinions and arguments of others combine to guarantee the quality of the research (Lilja et al., 2011).

For this research, a group of 17 experts was selected on the basis of their international scope and the following characteristics:

1. Academic experts (international researchers) with prestigious publications in the field of tourism economics and tourism sustainability (9 of the 17 invited experts);
2. Qualified representatives of national and international public administrations with responsibility in the management of tourism destinations (3 of the 17 invited experts);
3. Members of institutions and/or business organizations linked to tourism activity (3 of the 17 invited experts); and
4. Professionals from the field of consultancy with experience in tourism planning and management (2 of the 17 invited experts).

The empirical work was carried out in March and April of 2022. A sine qua non condition was established in the existence of a close link between the professional, institutional, and/or research activity of the experts invited to collaborate, and the topics under study (with a minimum experience of five years, as established by Donohoe (2011). Contact was via email, SMS, or telephone, clearly outlining the objectives of the study, as well as to why opinion of the expert was important. The anonymous nature of the responses was also explained.

Statistical processing of the questionnaires was carried out using SPSS (Statistical Package for the Social Sciences) software. The major statistics used in Delphi studies are measures of central tendency and level of dispersion—standard deviation and interquartile range (IQR)—in order to present information concerning the collective judgements of respondents

(36). The use of the median and mode is favored (in the literature, the use of median scores, based on Likert-type scales, is strongly favored) (37–39). As stated by Donohoe and Needham (2009), consensus is achieved when there is agreement among 60 per cent of the members of the panel. In addition, the interquartile range must not exceed 10 per cent above or below the mid-point. Following these recommendations, both criteria are present in the analysis of the results. The level of consensus reached is established as follows:

- High consensus: 60 per cent of the experts (or more) have reached the median value;
- Fair consensus: between 30 per cent and 59.99 per cent of the experts reached the median value; and
- Low consensus: less than 30 per cent of the experts reached the median value.

Moreover, in relation to the IQR, it is estimated that:

- High degree of consensus: the IQR does not exceed 0.7 (which would be 10 per cent of the 7 points of the Likert scale used);
- Considerable consensus: the IQR reaches values ranging from 0.7 to 1.5; and
- Not enough consensus: the value of IQR exceeds 1.5.

Results: A Proposal of Indicators to Measure Tourists' Sustainable Intelligence

In line with the classical approach of the Delphi methodology, it was decided that the first questionnaire would have open and semi open-ended questions (Garatti et al., 2002; Mitchell, 1991). A first testing round with experts was held in order to present the results of the previous research, which creates a context for this study, as it presents for the first time the notion of tourists' sustainable intelligence (López-Sánchez & Pulido-Fernández, 2016). Based on these premises, open and semi

Table 9.1 Dimensions of tourists' sustainable intelligence

Dimensions of tourists' sustainable intelligence	**Dimension 1.** Tourists' knowledge of the meaning of sustainability and sustainable tourism **Dimension 2.** Tourists' attitude in relation to sustainability and sustainable tourism **Dimension 3.** Sustainable Tourist Behavior **Dimension 4.** Sustainable tourist consumption **Dimension 5.** Tourists' economic commitment to the sustainability of destinations

open-ended questions were posed to the experts to determine the main dimensions that should shape the indicators measuring the sustainable intelligence of tourists. The initial proposals were based on an in-depth review of the literature on what it means for a tourist to be sustainable or pro-sustainable (Budeanu, 2007; Budeanu et al., 2016; Dolnicar, 2004; Fennell, 2021; Passafaro et al., 2015, among others). Table 9.1 shows the established dimensions:

Prior to sending out the questionnaire for the first round (once the dimensions had been verified), a pre-test was carried out with a small group of experts to verify that the wording was clear and comprehensible, that the judgements included did not condition the opinions of the experts, and that there were no misleading statements. In this way, the content of the questionnaire was validated before the start of the official rounds. In addition, the experts were also asked to provide their own contributions (to include, modify, or omit any global indicators) to be discussed in the second round. As a result, some indicators that might give rise to divergent interpretations were reworded to improve their comprehension. Experts were asked to rate the categorized responses on a scale from 1 to 7, with 1 being "very irrelevant" and 7 being "extremely relevant".

With regard to the final content of the questionnaire, after identifying the main dimensions (knowledge, attitude, behavior, consumption, and economic commitment), a series of indirect indicators were established to quantify the sustainable intelligence of tourists. Therefore, the objective, after validating the main dimensions, was to define the principal (generic) overall indicators to be included in these dimensions. It should be noted that, as mentioned above, in the following phases of this research, the plan is to evaluate, design, and include, for each of these general indicators, a series of specific indicators that provide more detailed content.

Table 9.2 shows in detail, for each dimension, the five global indicators included by the experts, as well as the values, for each of the three rounds, of the mean (Q2) and the IQR (Q3-Q1). The basis for further refinement of the set of indicators of tourists' sustainable intelligence is established as follows:

- **Dimension 1. Tourists' knowledge of the meaning and implications of sustainability and sustainable tourism:** The aim is to ascertain tourists' level of knowledge and understanding about the meaning of sustainability (as a global concept, not exclusively applied to tourism activity) and sustainable tourism. In addition, indicators are included to assess whether, in their expert opinion, tourists have sufficient knowledge and judgement to assess or measure the sustainability of destinations. In other words, does the destination effectively communicate its commitment to sustainability? Finally, it is important to know whether and to what extent tourists are included as agents in the destination, responsible for the progress of the destination or its setbacks in terms of sustainability. Future specific indicators (which will define the content of global indicators) will explore in greater depth issues related to the communication of the destination in terms of sustainability and the involvement of tourists as critical and responsible agents in the sustainability of destinations.
- **Dimension 2. Tourists' attitude in relation to sustainability and sustainable tourism**: The aim is to measure the declared (unimplemented) commitment of tourists in relation to the importance of advancing sustainability (at a global level) and the sustainability of tourist destinations, in particular. Experts believe it is important to know the capacity of tourists to identify aspects related to sustainable tourist behavior and their declared commitment to favor such behavior. In addition, it includes the motivations of tourists to manifest this level of commitment.

Table 9.2 Values of the Delphi questionnaire about "indicators of tourists' sustainable intelligence"

	Validity		Importance (1 to 7)					
			Round 1		Round 2		Round 3	
	Yes	No	Q2	Q3–Q1	Q2	Q3–Q1	Q2	Q3–Q1
Dimension 1. Tourists' knowledge of the meaning and implications of sustainability and sustainable tourism								
GLOBAL INDICATOR 1. Tourists' level of knowledge about the meaning of sustainability	X		6	1.5	6	1	7	1
GLOBAL INDICATOR 2. Tourists' level of knowledge about the meaning of sustainable tourism	X		5	1.5	6	1	7	0.5
GLOBAL INDICATOR 3. Tourists' capacity (elements of judgement) to evaluate the sustainability of a destination	X		6	1	7	1	7	0.5
GLOBAL INDICATOR 4. Tourists' capacity to identify all the territorial agents responsible for destination sustainability	X		5	2	5	1.5	6	1
GLOBAL INDICATOR 5. Tourists' capacity to identify themselves as a crucial agent in destination sustainability	X		6	1.5	6	1	6	0.5
Dimension 2. Tourists' attitude in relation to sustainability and sustainable tourism								
GLOBAL INDICATOR 1. Tourists' position in relation to the importance of advancing sustainability globally	X		6	2	7	1	7	0.5
GLOBAL INDICATOR 2. Tourists' position in relation to the importance of advancing the sustainability of tourist destinations	X		7	2	7	0.5	7	0
GLOBAL INDICATOR. 3 Tourists' capacity to identify aspects related to sustainable tourist behavior	X		6	2	6	1	7	1
GLOBAL INDICATOR 4. Tourists' declared commitment to identify aspects related to sustainable tourist behavior	X		6	1	6	0.5	7	0.5
GLOBAL INDICATOR 5. Tourists' motivations to manifest said level of commitment to aspects related to sustainable tourist behavior	X		5	2	6	1	6	0.5
Dimension 3. Sustainable tourist behavior								

(continued)

Table 9.2 (continued)

	Validity		Importance (1 to 7)					
			Round 1		Round 2		Round 3	
	Yes	No	Q2	Q3–Q1	Q2	Q3–Q1	Q2	Q3–Q1
GLOBAL INDICATOR 1. Tourists' level of knowledge about the meaning and implications of sustainable tourism	X		6	2	6	1	6	0.5
GLOBAL INDICATOR 2. Tourists' level of knowledge about the meaning of sustainable tourist behavior	X		6	1.5	6	0.5	6	0.25
GLOBAL INDICATOR 3. Importance attached by tourists to different aspects related to sustainable tourist behavior	X		5	2	6	6	6	0.5
GLOBAL INDICATOR 4. Tourists' level of compliance in the destination with aspects linked to sustainable tourist behavior	X		6	1	7	1	7	0.25
GLOBAL INDICATOR 5. Tourists' reasons to comply (or not) with the destination with aspects linked to sustainable tourist behavior	X		6	2	6	1	7	1
Dimension 4. Sustainable tourist consumption								
GLOBAL INDICATOR 1. Tourists' level of knowledge about the meaning of sustainable consumption	X		5	2	5	1	6	0.5
GLOBAL INDICATOR 2. Tourists' level of knowledge about the meaning of sustainable tourist consumption	X		6	1	6	0.5	7	0.75
GLOBAL INDICATOR 3. Tourists' level of sustainable consumption during their stay	X		6	0.5	7	1	7	0.5
GLOBAL INDICATOR 4. Level of sustainable consumption before their stay (costs related to their future trip)	X		5	2	6	1	6	0.25
GLOBAL INDICATOR 5. Tourists' reasons to carry out (or not) sustainable consumption	X		6	2	6	0.5	7	0.75
Dimension 5. Tourists' economic commitment to the sustainability of destinations								

(*continued*)

Table 9.2 (continued)

	Validity		Importance (1 to 7)					
			Round 1		Round 2		Round 3	
	Yes	No	Q2	Q3–Q1	Q2	Q3–Q1	Q2	Q3–Q1
GLOBAL INDICATOR 1. Reasons given by tourists for accepting (or not) that sustainable destinations might be more expensive (either as a differentiation strategy or derived from higher costs through their commitment to sustainability)	X		6	2	6	0.5	7	0.5
GLOBAL INDICATOR 2. Tourists' willingness to pay more for a sustainable tourist destination	X		6	0.5	7	1	7	0.5
GLOBAL INDICATOR 3. Tourists' motivation to pay more (or not) for a sustainable tourist destination	X		7	1	7	0.75	7	0.25
GLOBAL INDICATOR 4. Tourists' willingness to pay a tax ("eco tax") to improve the sustainability of destinations	X		6	1	6	0	7	0.5
GLOBAL INDICATOR 5. Tourists' motivation to pay (or not) a tax ("eco tax") to improve the sustainability of destinations	X		6	2	6	0	7	0.5

Source: Authors' own elaboration

- **Dimension 3. Sustainable tourist behavior:** The aim is to ascertain not only the declared intentions, but the actual behavior of tourists, their motivations, and the causes of their behaviors. It is important to understand the level of knowledge of the tourist with respect to the meaning and implications of sustainable behavior (as a lifestyle) and sustainable tourist behavior. Having evaluated levels of knowledge about these issues, the importance that tourists attach to these aspects related to sustainable tourist behavior is measured.
- **Dimension 4. Sustainable tourist consumption:** The main difference between dimensions 3 and 4 is the economic factor, since, according to the experts, it is important to distinguish between

behavior (without economic implications for the tourist) and consumption choices. This dimension allows us to determine tourists' level of knowledge about the meaning and significance of sustainable consumption in general and in more specific terms. Sustainable tourist consumption is measured and quantified before (through expenses related to their future trip) and during their stay, along with the main reasons given by tourists for engaging (or not) in sustainable consumption.

– **Dimension 5. Tourists' economic commitment to the sustainability of destinations:** This dimension includes global indicators related to the reasons given by tourists for accepting (or not) that sustainable destinations might be more expensive (either as a differentiation strategy or derived from higher costs through their commitment to sustainability). It also evaluates tourists' willingness to pay more for a sustainable tourist destination and the reasons for their choices. Furthermore, it ascertains tourists' willingness to pay a tax ("eco tax") to improve destination sustainability and the reasons for this willingness to pay or not pay.

Discussion and Conclusions

A city is not smart because it uses technology. A city is smart because it uses technology (and other management strategies and tools) to make the lives of its citizens better and to optimize efficiency and effectiveness in the pursuit of competitiveness and sustainability. However, there remains insufficient research on citizen participation in the context of smart cities and, consequently, there are still many questions to be answered regarding the participation, power, and impact of tourists on smart destinations.

Reviewing the scientific literature, it is not difficult to find relevant studies that address and measure, using different methodologies, aspects related to tourists and the sustainability of destinations, from a variety of perspectives. For example, there are studies that analyze tourists' knowledge of sustainability (López-Sánchez & Pulido-Fernández, 2016), the attitude–behavior gap in sustainable tourism (Juvan & Dolnicar, 2014), tourists' willingness to pay for more sustainable destinations (López-Sánchez & Pulido-Fernández, 2014, 2017), sustainability and tourist taxes (Durán-Román et al., 2021), and sustainable tourism indicators in relation to the Sustainable Development Goals (Rasoolimanesh et al., 2020), etc.

This study lays the foundations for a holistic approach that brings together all of these facets to build a single tool that can be used to better understand the sustainable intelligence of tourists, that is, the knowledge, attitudes, consumption, and economic commitment of tourists with respect to the sustainability of destinations.

One of the highlights of this study is that each dimension includes global indicators that provide analysis from the general to the specific. In this sense, this project contributes knowledge in relation to individual versus societal consequences, immediate versus deferred consequences, and local versus global consequences. This set of indicators allows us, for example, to ascertain whether pro-sustainable or responsible behavior in the daily life of citizens is transferred to their consumption as tourists and analyzes the causes and reasons for possible misalignments. These indicators provide insight into the level of theoretical knowledge of sustainability and its practical implications.

If smart cities are a sine qua non condition for the existence of smart destinations (or smart sustainable destinations), does an individual with a high level of sustainable intelligence translate into a tourist with the same level of sustainable intelligence? Empirical experience tells us that this is not necessarily the case (Hibbert et al., 2013; Higham et al., 2016; Juvan & Dolnicar, 2014). There are multiple reasons for the contradictory nature between pro-sustainable concerns and consumption decisions in everyday life and in tourist contexts.

This chapter provides the necessary theoretical basis for further research in order to develop a new system of indicators capable of measuring the sustainable intelligence of tourists. A smart city and a smart destination must attract and satisfy smart citizens and smart tourists and sustainability must be the cross-cutting axis that sustains and guides the process. Creating tools to more clearly understand the different variables that define and condition sustainability for tourists is vital in order to help us all transition into a sustainable future.

References

Adler, M., & Ziglio, E. (1996). *Gazing into the oracle: The Delphi method and its application to social policy and public health*. Jessica Kingsley Publishers.

Ahvenniemi, H., Huovila, A., Pinto-Seppä, I., & Airaksinen, M. (2017). What are the differences between sustainable and smart cities? *Cities, 60*, 234–245. https://doi.org/10.1016/j.cities.2016.09.009

Benevolo, C., Dameri, R. P., & D'Auria, B. (2016). Smart mobility in Smart City action taxonomy, ICT intensity and public benefit. In T. Torre, A. Braccini, & R. Spinelli (Eds.), *Empowering organizations. Lecture notes in information systems and organisation* (Vol. 11, pp. 13–28). Springer. https://doi.org/10.1007/978-3-319-23784-8_2

Bibri, S. E., & Krogstie, J. (2017). Smart sustainable cities of the future: An extensive interdisciplinary literature review. *Sustainable Cities and Society, 31*, 183–212. https://doi.org/10.1016/j.scs.2017.02.016

Budeanu, A. (2007). Sustainable tourist behaviour—A discussion of opportunities for change. *International Journal of Consumer Studies, 31*(5), 499–508. https://doi.org/10.1111/j.1470-6431.2007.00606.x

Budeanu, A., Miller, G., Moscardo, G., & Ooi, C. S. (2016). Sustainable tourism, progress, challenges and opportunities: An introduction. *Journal of Cleaner Production, 111*, 285–294. Elsevier Ltd.. https://doi.org/10.1016/j.jclepro.2015.10.027

Buhalis, D., & Amaranggana, A. (2014, January 21–24). Smart tourism destinations. In *Information and communication technologies in tourism 2014 proceedings of the international conference in dublin* (pp. 553–565). https://doi.org/10.1007/978-3-319-03973-2.

Chourabi, H., Nam, T., Walker, S., Gil-Garcia, J. R., Mellouli, S., Nahon, K., Pardo, T. A., & Scholl, H. J. (2012). Understanding smart cities: An integrative framework. *Proceedings of the Annual Hawaii International Conference on System Sciences, 2289–2297*. https://doi.org/10.1109/HICSS.2012.615

Dalkey, N., Brown, B., & Cochran, S. (1972). *La Prévision à long terme par la méthode Delphi*. Dunod.

Dalkey, N., & Helmer, O. (1963). An experimental application of the Delphi method to the use of experts. *Management Science, 9*(3), 458–467.

Datta, A. (2015). New urban utopias of postcolonial India: 'Entrepreneurial urbanization' in Dholera smart city, Gujarat. *Dialogues in Human Geography, 5*(1), 3–22. https://doi.org/10.1177/2043820614565748

Dolnicar, S. (2004). Insights into sustainable tourists in Austria: A data-based a priori segmentation approach. *Journal of Sustainable Tourism, 12*(3), 209–218. https://doi.org/10.1080/09669580408667234

Donohoe, H. M. (2011). Defining culturally sensitive ecotourism: A Delphi consensus. *Current Issues in Tourism, 14*(1), 27–45. https://doi.org/10.1080/13683500903440689

Donohoe, H. M., & Needham, R. D. (2009). Moving best practice forward: Delphi characteristics, advantages, potential problems, and solutions. *International Journal of Tourism Research, 11*(5), 415–437.

Durán-Román, J. L., Cárdenas-García, P. J., & Pulido-Fernández, J. I. (2021). Tourists' willingness to pay to improve sustainability and experience at destination. *Journal of Destination Marketing and Management, 19*(100540). https://doi.org/10.1016/j.jdmm.2020.100540

Femenia-Serra, F., Neuhofer, B., & Ivars-Baidal, J. A. (2018). Towards a conceptualisation of smart tourists and their role within the smart destination scenario. *The Service Industries Journal*, 109–133.

Fennell, D. A. (2021). Technology and the sustainable tourist in the new age of disruption. *Journal of Sustainable Tourism, 29*(5), 767–773. https://doi.org/10.1080/09669582.2020.1769639

Fletcher, R., Murray Mas, I., Blanco-Romero, A., & Blázquez-Salom, M. (2019). Tourism and degrowth: An emerging agenda for research and praxis. *Journal of Sustainable Tourism, 27*(12), 1745–1763. https://doi.org/10.1080/09669582.2019.1679822

Garatti, M., Costa, R., Reghizzi, S. C., & Rohou, E. (2002). The impact of alias analysis on VLIW scheduling. In *Lecture notes in computer science (including subseries lecture notes in artificial intelligence and lecture notes in bioinformatics), 2327 LNCS* (pp. 93–105). https://doi.org/10.1007/3-540-47847-7_10.

Granier, B., & Kudo, H. (2016). How are citizens involved in smart cities? Analysing citizen participation in Japanese "smart communities". *Information Polity, 21*(1), 61–76. https://doi.org/10.3233/IP-150367

Green, H., Hunter, C., & Moore, B. (1990). Assessing the environmental impact of tourism development. Use of the Delphi technique. *Tourism Management, 11*(2), 111–120. https://doi.org/10.1016/0261-5177(90)90026-6

Guo, Y., Liu, H., & Chai, Y. (2014). The embedding convergence of smart cities and tourism internet of things in China: An advance perspective. *Advances in Hospitality and Tourism Research, 2*(1), 54–69. https://dergipark.org.tr/en/pub/ahtr/issue/32308/359048

Gupta, K., & Hall, R. P. (2017). The Indian perspective of smart cities. In *2017 Smart cities symposium prague, SCSP 2017—IEEE proceedings*. https://doi.org/10.1109/SCSP.2017.7973837.

Haarstad, H. (2017). Constructing the sustainable city: Examining the role of sustainability in the 'smart city' discourse. *Journal of Environmental Policy and Planning, 19*(4), 423–437. https://doi.org/10.1080/1523908X.2016.1245610

Hibbert, J. F., Dickinson, J. E., Gössling, S., & Curtin, S. (2013). Identity and tourism mobility: An exploration of the attitude-behaviour gap. *Journal of Sustainable Tourism, 21*(7), 999–1016. https://doi.org/10.1080/09669582.2013.826232

Higham, J., Reis, A., & Cohen, S. A. (2016). Australian climate concern and the 'attitude–behaviour gap'. *Current Issues in Tourism, 19*(4), 338–354. https://doi.org/10.1080/13683500.2014.1002456

Ivars-Baidal, J. A., Celdrán-Bernabeu, M. A., Mazón, J. N., & Perles-Ivars, Á. F. (2019). Smart destinations and the evolution of ICTs: A new scenario for destination management? *Current Issues in Tourism, 22*(13), 1581–1600. https://doi.org/10.1080/13683500.2017.1388771

Jasrotia, A., & Gangotia, A. (2018). Smart cities to smart tourism destinations: A review paper. *Journal of Tourism Intelligence and Smartness, 1*(1), 47–56. https://dergipark.org.tr/en/download/article-file/530727

Juvan, E., & Dolnicar, S. (2014). The attitude-behaviour gap in sustainable tourism. *Annals of Tourism Research, 48*, 76–95. https://doi.org/10.1016/j.annals.2014.05.012

Kaynak, E., & Macaulay, J. A. (1984). The Delphi technique in the measurement of tourism market potential. The case of Nova Scotia. *Tourism Management, 5*(2), 87–101. https://doi.org/10.1016/0261-5177(84)90056-6

Landeta, J. (2006). Current validity of the Delphi method in social sciences. *Technological Forecasting and Social Change, 73*(5), 467–482. https://doi.org/10.1016/j.techfore.2005.09.002

Lilja, K. K., Laakso, K., & Palomki, J. (2011). Using the delphi method. In *PICMET: Portland international center for management of engineering and technology, proceedings, 1999.*

Lin, V. S., & Song, H. (2015). A review of Delphi forecasting research in tourism. *Current Issues in Tourism, 18*(12), 1099–1131. https://doi.org/10.1080/13683500.2014.967187

Linstone, H. A., & Turoff, M. (2002). The delphi method: Techniques and applications. In *Skeletal radiology*. Addison-We. https://doi.org/10.1007/s00256-011-1145-z.

López-Sánchez, Y., & Pulido-Fernández, J. I. (2014). Factors determining the willingness to pay for a more sustainable destination. The case of the Western Costa del Sol. *Cuadernos de Turismo, 33*.

López-Sánchez, Y., & Pulido-Fernández, J. I. (2016). In search of the pro-sustainable tourist: A segmentation based on the tourist "sustainable intelligence". *Tourism Management Perspectives, 17*. https://doi.org/10.1016/j.tmp.2015.12.003

López-Sánchez, Y., & Pulido-Fernández, J. I. (2017). Factors influencing the willingness to pay for sustainable tourism: A case of mass tourism destinations. *International Journal of Sustainable Development and World Ecology, 24*(3). https://doi.org/10.1080/13504509.2016.1203372

Mitchell, V. W. (1991). The Delphi technique: An exposition and application. *Technology Analysis & Strategic Management, 3*(4), 333–358. https://doi.org/10.1080/09537329108524065

OECD. (2020). Smart cities and inclusive growth. In *Smart cities and inclusive growth © OECD 2020, typology of smart cities* (pp. 1–59).

Passafaro, P., Cini, F., Boi, L., D'angelo, M., Heering, M. S., Luchetti, L., Mancini, A., Martemucci, V., Pacella, G., Patrizi, F., Sassu, F., & Triolo, M. (2015). The "sustainable tourist": Values, attitudes, and personality traits. *Tourism and Hospitality Research, 15*(4), 225–239. https://doi.org/10.1177/1467358415576086

Perles-Ribes, J. F., & Ivars-Baidal, J. (2018). Smart sustainability: A new perspective in the sustainable tourism debate. *Investigaciones Regionales, 42*, 151–170.

Perles-Ribes, J. F., & Ramón-Rodríguez, A. B. (2017). Calp, hacia la sostenibilidad a través de la inteligencia. In *Congresos—Seminario Destinos Turisticos Inteligentes 2017—Libro de Actas* (pp. 71–95). https://doi.org/10.14198/destinos-turisticos-inteligentes.2017.04.

Pill, J. (1971). The Delphi method: Substance, context, a critique and an annotated bibliography. *Socio-Economic Planning Sciences, 5*(1), 57–71. https://doi.org/10.1016/0038-0121(71)90041-3

Pulido-Fernández, J. I. (2005). *Criterios para una política turística sostenible en los parques naturales de Andalucía*. Doctoral dissertation, University of Jaén.

Rasoolimanesh, S. M., Ramakrishna, S., Hall, C. M., Esfandiar, K., & Seyfi, S. (2020). A systematic scoping review of sustainable tourism indicators in relation to the sustainable development goals. *Journal of Sustainable Tourism*, 1–21. https://doi.org/10.1080/09669582.2020.1775621

Silva, B. N., Khan, M., & Han, K. (2018). Towards sustainable smart cities: A review of trends, architectures, components, and open challenges in smart cities. *Sustainable Cities and Society, 38*, 697–713. https://doi.org/10.1016/j.scs.2018.01.053

Walter, C., Gianluca, M., & Alberto, S. (2015). Citizen's engagement and value co-production in smart and sustainable cities. *International Conference on Public Policy*, 1–16. http://www.icpublicpolicy.org/conference/file/reponse/1433973333.pdf

Williams, A. M., Rodriguez, I., & Makkonen, T. (2020). Innovation and smart destinations: Critical insights. *Annals of Tourism Research, 83*, 1–10. https://doi.org/10.1016/j.annals.2020.102930

Part IV

Sustainable Tourism: Policies and Practices

10

Tourism Views and Experiences of Micro- and Small Enterprises in Northern Finland and Russian Karelia during the COVID-19 Pandemic: A Cross-Border Operations Perspective

Katariina Ala-Rämi and Kyllikki Taipale-Erävala

Introduction

Tourism is a growing industry and significant generator of new jobs, which can benefit declining regions and play an important role in rural areas, where distinctive social and cultural heritage landscapes not only attract tourists, but also offer alternatives to mass tourism destinations (Gavrila-Paven et al., 2015; Irvine & Anderson, 2007). For this study, we define sustainable tourism as those activities that meet the needs of tourists and host regions while protecting and enhancing opportunities for

K. Ala-Rämi (✉)
University of Oulu, Oulu, Finland
e-mail: katariina.ala-rami@oulu.fi

K. Taipale-Erävala
University of Eastern Finland, Joensuu, Finland
e-mail: kyllikki.taipale-eravala@uef.fi

T. Walker et al. (eds.), *Sustainable Tourism*,
https://doi.org/10.1007/978-3-031-43528-7_10

the future.[1] The World Tourism Organization defines sustainable tourism as: "Tourism that takes full account of its current and future economic, social, and environmental impacts, addressing the needs of visitors, the industry, the environment and host communities" (UN, 2022). This definition includes the idea that tourism should not be curtailed globally because it is important to consider it in the larger context of people's well-being and sustainability. The negative effects of tourist activities can be reduced by favoring destinations which require less travel and that are not already overcrowded, but this can also be achieved if increasing tourism does not lead to building large tourist centers, instead respecting indigenous nature, culture, and inhabitants of a given region (Nicholls, 2006). Tourism in northern regions provides a source of livelihood to permanent residents. Thus, supporting this opportunity creates an ecologically, economically, and socially sustainable counterforce to an increasing trend of movement toward growth centers (Irvine & Anderson, 2007). Northern Europe has a long history of smooth cross-border operations, offering easy cross-border tourism experiences that have been an important part of regional development processes and economic life. Regular cross-border tourism supports not only economic life, but also political and cultural integration (e.g., Stoffelen & Vanneste, 2017). To promote the sustainable growth and renewal of the tourism sector, the strategy defines four key priorities: supporting activities that foster sustainable development, responding to digital change, improving accessibility, and developing an operating environment that promotes competitiveness (OECD, 2020a).

The Tornio-Haparanda area, which is located along the Finnish-Swedish border, constitutes a cultural region with a long history and is often studied as a good example of a strong cross-border region (Prokkola, 2007), where cooperation between sister cities and their inhabitants has been actively developed, deepened, and increased over several decades. On the contrary, approximately 300 kilometers east of the Swedish border is the Russian border, with Karelian descendants residing on both sides; here, political tension and strong control exist, resulting in

[1] https://www.unwto.org/sustainable-development.

cross-border tourism being quite underutilized (Izotov & Laine, 2013, Manakov et al., 2020). It is important to note here that these three countries share a long history, both geographically and culturally. Located between Sweden and Russia, Finland was part of Sweden until 1809, when it was taken over by Russia until its declaration of independence in 1917.

This chapter explores the multidimensional tourism development possibilities and challenges in a strong cross-border region; hence, the study region consists of two Finnish counties, namely, Northern Ostrobothnia and Kainuu, and a third region from the Russian side, Russian Karelia (Fig. 10.1). Our study, conducted before the invasion of Ukraine, provides a novel understanding of borderland tourism possibilities in less-favored regions in the context of sustainability. Cross-border tourism research covers mostly those borderlands that have lively tourism activities, whereas our political, economic, sociocultural, technological, environmental, and legal (PESTEL) analysis-based study focuses on the region where sustainable tourism potential is highly untapped. The research was carried out during the cross-border project, which provided rich materials and an understanding of the complexity of the issue.

We suggest that all PESTEL factors need to be considered in sustainable tourism businesses. On both sides of their borders, these regions offer unique experiences and quality-driven outdoor activities that are ecologically safer and more sustainable than in the polar regions (see Narvik University College, 2010). These are increasingly valued aspects of tourism based on sustainability and digitalization (OECD, 2020a). However, to utilize sustainable tourism opportunities, there is a need to observe challenges that include not only linguistic and cultural issues, but also economic and political aspects that a strong EU-Russia, external, 1340-kilometer-long border creates. This chapter is the result of an ongoing need to explore better tourism possibilities and their challenges, especially in the northern parts of these sparsely populated rural borderland regions, where the growth of the tourism sector could increase the regions' vitality, which, in turn, can be utilized in sustainable ways.

In the "Methodology" section of this chapter, we discuss our methodology before explaining the PESTEL analysis tool we have used in the "PESTEL Analysis" section. The next section, "Tourism in Northern

Fig. 10.1 The study region: Northern Ostrobothnia (Oulu Region), Kainuu, and Republic of Karelia

Ostrobotnia, Kainuu, and Russian Karelia" provides an overview of the tourism industry in Northern Ostrobothnia, Kainuu, and Russian Karelia, while the following section, "Cross-border Tourism and Livelihood in Sparsely Populated Areas (SPAs)" looks at cross-border tourism in sparsely populated areas (SPAs). In the next section "Results of the PESTEL Analysis," we examine our project's results before providing our concluding discussion in section "Discussion and Conclusion."

Methodology

This chapter explores the possibilities and challenges for the growth of tourism businesses in a designated area, which consists of three regions along the Finnish and Russian borders. It offers understanding and experiences to broader SPAs and other cross-border regions experiencing challenges in employment and livelihood. Attracting international tourists requires specific service skills, such as the ability to speak foreign languages (at least English) and cultural understanding, but tourism companies in these regions are mostly small and have limited resources. Thus, they often lack international service development competence and knowledge of international tourism standards and demands (Taipale-Erävala et al., 2014). They also have limited experience coordinating cross-border activities and functions. The growth of the tourism industry could create new income streams that would strengthen the livelihoods of these less-favored, remote, and SPAs.

We chose a case study approach to clarify existing tourism business possibilities. Our study focuses on Finnish-Russian tourism cooperation between small and medium enterprises (SMEs), aiming to support tourism enterprises in initiating cross-border business cooperation in Northern Ostrobothnia, Kainuu, and Russian Karelia (see Fig. 10.1).

The collected data is based on primary data sources, such as notes, memos, discussions, records, and entries, made within the TourSME project. We collected the data and observations when conducting the research project in Finland and Russia from tourism enterprises intensively involved in cross-border activities and tourism-related stakeholders, as well as written national, regional, and local tourism materials. The secondary data sources are publicly available Finnish and Russian electronic tourism-related sources (newspapers, web pages, social media, etc.). Because the research was carried out in the context of a larger project in the case study's location, it allowed for data to be collected in a rich setting, where we experienced the region and its actors with all senses and were able to collaborate with not only the enterprises involved in project's activities, but also to carry out project activities and operations with five project partners: three Finnish and two Russian partners. During the preparation phase, all the project partners collaborated to create the motivation mapping of the challenges, which helped in creating a sketch for PESTEL analysis. The authors represent the research project's lead partner, the University of Oulu (FI) along with the Oulu University of Applied Sciences (FI), which are in Northern Ostrobothnia, and the third Finnish partner, KAMK—University of Applied Science (FI), which is in Kainuu. The Russian partners were located in the City of Kostamusha (RU) and Karelian Research Center (RU) from Petrozavodsk.

Additionally, we have both personal and practical experience in the Finnish and Russian tourism industries, having worked in the field for roughly 20 years. Experience and tourism knowledge were utilized alongside the primary and secondary data according to PESTEL analysis factors, which are described below. The PESTEL analysis was used to evaluate the factors affecting tourism activities in less-favored and remote SPAs. The PESTEL categories and factors are presented in Finnish and Russian cross-border contexts, bearing in mind the business situations in which the tourism companies operate.

PESTEL Analysis

PESTEL analysis is considered a strategic planning method for companies targeted at international markets by helping assess the behavior of tourism demands (Gregorić, 2014). An advantage of the PESTEL analysis is that it is a simple framework to use, helps one to understand wider business environments, and enables broader external and strategic thinking. However, it has some disadvantages, which are mainly connected to data analysis, such as capturing extensive data, simplifying data, or using data based on assumptions (CIPD Community, 2021). To avoid these disadvantages, we paid additional attention to collecting data and processing it. The PESTEL analysis considers the effects or potential impacts of external elements of the business environment in any industry, which can be classified into six categories (or more depending on the variant of the tool used). As mentioned above, for the purpose of this study, we used these six categories as a framework: political, economic, sociocultural, technological, environmental, and legal.

Political issues were evaluated considering factors related to Finnish and Russian national or international political situations, political stability, policy priorities, and the introduction of new policy initiatives related to the tourism business. Economic factors referred to global and national economies, as well as financial issues and markets. Sociocultural factors explored include the impact of demographic changes and changes in customer behavior and other personal behavior that enable business opportunities in SPAs. Changes in lifestyle and attitudes toward work and leisure were also considered sociocultural factors. Technological factors examined include development trends in technology, either in the general form of IT development or digitalization, as related to advancing the tourism industry. Environmental factors considered relate to the natural environment and ecological concerns. Finally, legal issues, which involve possible changes in national and/or international laws, as well as regulations originating from the domestic and international markets, were considered.

Tourism in Northern Ostrobotnia, Kainuu, and Russian Karelia

The regions we consider have narrow and vulnerable business environments and are among the most sparsely populated in Europe, with a population density of only 4.9 inhabitants/km^2 (Carbone, 2018). The midsummer sun, the northern lights, and reindeer remain typical tourist attractions associated specifically with Lapland (Maher et al., 2014), even though such attractions can be experienced throughout the studied region. The COVID-19 pandemic resulted in a rapid decrease in overnight stays in Northern Ostrobotnia, Kainuu, and Russian Karelia during the spring of 2020 (Ragulina, 2020; Visit Finland, 2021). However, by July and August 2020, overnight stays in these regions increased and were even higher during the summer of 2021 than in the previous year because of pandemic restrictions that increased domestic tourism with travelers avoiding going abroad (Visit Finland, 2021). Tourism has a positive influence on other sectors and is therefore recognized as one of the main sources of new business development in the studied region, having some well-known tourist destinations, such as Ruka and Vuokatti. In light of the numerous natural attractions and seasonal experiences, such as snowfalls and northern lights, most of this borderland region's possibilities are yet underutilized. Hence, new cross-border tourism destinations are currently being fostered by local micro, small, and medium enterprises (SMEs).

According to information provided from our project partner in the city of Kostomuksha, cross-border tourism is also vital for Russian Karelia. In this region, the great majority of tourists, including transit tourists,[2] come from Finland. According to the Russian Customs Service (Y. Ragulina, personal communication, August 31, 2020), about 275,000 people entered and left the Kostomuksha territory in 2019. This information is based on hotel owners' notifications; the total number of overnight stays before the pandemic was about 7000 during the first three months of 2020. However, when the pandemic began, the number of overnight stays dropped to 1000 by the end of August 2020 (Y. Ragulina, personal communication, August 31, 2020). This represents a critical change in demand, and the usage of accommodation capacity fell to less than one-tenth.

Tourism is highly significant in these remote areas because it not only brings opportunities to traditional tourism businesses but also adds subsidiary trade, that is, farmers and fishers and more customers to shops and health and wellness sector. Even though there is no unambiguous way to calculate the economic impacts of tourism for the region, tourism has a strong multiplicative effect, increasing local income and creating employment opportunities for other businesses such as groceries and well-being (Hakkarainen & Tuulentie, 2008; Makkonen et al., 2018; Saarinen, 2007). Additionally, tourism firms possess additional part-time business opportunities, for example, in small farms that cannot compete with the productivity of those that are centrally located and in more favorable circumstances (Gavrila-Paven et al., 2015).

There exist unique cultures, natural attractions, historical values, languages, and many other tourism-related strengths (Irvine & Anderson,

[2] Transit tourists are people stopping over in a country on their way to another destination.

2007) on both sides of the border that are not widely recognized as the tourism destinations; this is because the region is not easily reached and little promoted. These borderland regions offer a combination of two Karelian-based cultures that are distinct but also quite similar (Izotov & Laine, 2013), so encountering them both in the same trip creates an even more unique experience. Thus, observing the existing tourism potential and having knowledge of new business possibilities may provide a sideline for the locals, thereby supporting other traditional industries—such as agriculture, forestry, and reindeer husbandry—that have been in decline because of climate change.

As Timothy (1999) points out, these types of adjacent regions, especially those on national peripheries, as in our case study, have political boundaries that tend to hinder sustainable tourism development, despite their great potential. Thus, the development of the tourism industry is essential in these borderland SPAs for economic viability and to sustain natural and cultural resources without leading to the degradation of the sociocultural and natural environments.

We place an interest on highlighting tourism attractions based on the utilization of the unique features of the region (Irvine & Anderson, 2007) and, at the same time, on the protection of cultural and natural heritage while developing services and practices that can be sustained for future generations (Hakkarainen & Tuulentie, 2008). Because both the natural and cultural resources that underpin tourism endeavors are located across international borders, their ongoing sustainable use will require cross-border partnerships (Izotov & Laine, 2013; Timothy, 1999). Increasing tourism business requires the provision of more comprehensive services, which can only be attained with proper cooperation between national and international tourism enterprises. The Visit Finland strategy 2020 defines cross-cutting development themes as year-round activities, digitalization, and responsibility (Visit Finland, 2019). These themes are also recurrent aspects in developing tourism in the region (see Yurievna, 2011).

Because all the areas of our study region are very sparsely populated, the potential of attracting tourists would increase with collaboration across the administrative and even national borders. By focusing on their cross-border tourism potential, we are looking specifically at what kinds of business-related opportunities they may have and the challenges they may face. Currently, cooperation and cross-border information exchange between Finnish and Russian Karelian tourism companies are limited, which is understandable because of strong borders and very different languages. Their knowledge of each other's tourism offerings is also limited; people on both sides might not even recognize the international tourism potential and attractiveness within their own individual regions, nor the potential that would arise if they were to collaborate. On the other hand, differences, for example, in tourism-related skills, languages, culture, and working methods, offer a wide scale of possibilities for mutual learning (Makkonen et al., 2018).

Cross-border Tourism and Livelihood in Sparsely Populated Areas (SPAs)

This chapter focuses on the economic importance of tourism in SPAs that cross over a strong border and have strict control and regulation (see, e.g., Prokkola, 2007). The operational environment of tourism in our study region is focal because it consists of different counties and cross-border activities. The border between Finland and Russia is fascinating as a tourism destination, and although it has mostly been seen as a barrier because of strict regulation and language challenges, with its distinct nature and culture, it can be regarded as a destination in itself (Irvine & Anderson, 2007; Timothy, 1999).

The border of the European Union (EU) is also supported by the European Neighborhood Policy (ENP) to foster stability, security, and prosperity. Because one of the most important principles of the EU is the free movement of people and things within the area, the world's largest visa-free zone, Schengen Area,[3] which represents a zone where 26 European countries have abolished their internal borders for the free and unrestricted movement of people, Finland is also a gate to offering free access to all countries within the Schengen region. The Karelia CBC program focuses on cross-border collaboration with the aim of "creating

[3] https://www.schengenvisainfo.com/schengen-visa-countries-list/.

attractive regions for people and business" (Karelia CBC, 2021; Makkonen et al., 2018). Supporting collaboration over the border is not only a question of creating economic welfare for the region and its inhabitants, but also promoting political and cultural understanding that enables sustainable tourism (e.g., Stoffelen & Vanneste, 2017).

Tourism is a sector that provides opportunities in rural regions (Gavrila-Paven et al., 2015; Saarinen, 2007). However, tourism also accelerates climate change; in turn, tourism activities are impacted by such change (Saarinen & Tervo, 2006). However, the northern parts of Finland and Russian Karelia are suitable for common outdoor activities, such as hiking and fishing, offering several attractive and unique natural attractions, and are among the most unpolluted regions in Europe. Because of the limited number of actors and many tourism possibilities, cross-border cooperation offers a significant possibility to create attractive and sustainable tourism services and to gain more attention from global tourists (see Maher et al., 2014). For example, to keep this nature unspoiled in the future presents ecological and economic challenges. Public transit access within the region is one such challenge. Carpooling and other kinds of solutions should also be available to minimize the effects of private motoring. However, it is also important to note the distance between the origin and destination. When compared with the emissions of long-haul flights to well-known tourist attractions, relatively short drives do not place such a heavy burden on nature. With respect to tourism SMEs, strengthening tourism by using natural attributes would create jobs and economic stability (i.e., Makkonen et al., 2018; Saarinen, 2007).

To a certain degree, some regions depend on each other in terms of global, national, and local development, and tourism is one sector that is highly influenced by events and changes at all such levels. SPAs are more vulnerable because their industrial bases tend to be small. Tourism also has a strong influence over local communities, having environmental and economic effects, as well as affecting sociocultural structures, processes, and dynamics. Furthermore, tourism has a multiplying impact because it invites new investments, creates jobs and income, and makes local products more popular and more widely recognized (Gavrila-Paven et al., 2015; Saarinen, 2007).

Results of the PESTEL Analysis

The major goals of this study were to demonstrate the complexity of the factors that influence the development and operation of cross-border tourism businesses that provide tourist services and offer policy implications that could further strengthen tourism activities in a sustainable way (see Table 10.1).

Political Factors

Political stability, or the lack thereof, significantly influences the tourism industry (Makkonen et al., 2018; Webster & Ivanov, 2013). Until recently, both Finland and Russia have been politically stable, enabling a business-friendly environment for tourism SMEs. The European geopolitical situation has, however, changed because of Russia's attacks on Ukraine, which began in February 2022, narrowing the scope of options for tourism SMEs (UNWTO, 2022).

Finland and Russia differ in many ways. Finland is a small and stable republic with 5.5 million inhabitants (Statistics Finland, 2022), while Russia is a large federation with 145.9 million inhabitants (United Nations, 2021) that has moved in a more authoritarian direction during the rule of President Putin (Ulkopolitiikka, 2019). Moreover, Finland is a member of the EU, while Russia is not. Thus, Finland's EU membership and Russia's political regime mean that there are different rules and practices on both sides of the border, which already do and may present future obstacles to smooth cross-border tourism services. Additionally, EU Ukraine-related sanctions against Russia and Finland's NATO membership application process in the spring of 2022 increased political tensions from the Russian side, which also influenced cross-border tourism (POLITICO, 2022).

From the viewpoint of travelers, tourism is often connected to ideas of freedom and independence, so political and legislative factors may affect how tourists perceive cross-border international travel. The border between Finland and Russia is also a border between the EU and the Russian Federation. Finnish citizens are EU citizens with European

Table 10.1 PESTEL results for cross-border tourism advantages and disadvantages

Factor	Advantages		Disadvantages		Summary
	Finnish	*Russian*	*Finnish*	*Russian*	*Cross-border*
Political	• Business-friendly environments				• Business-friendly environments
	• Republic since 1917, member of European Union (EU)	• Large federation		• Federation and semi presidential republic since 1993, moved to more authoritarian direction during the rule of Putin	• Different political rules and practices on both sides of the border • Changed geopolitical situation since February 2022
	• Free movement for members of EU makes traveling across the borders in Europe easier	• Electronic visa for short-term and long-term tourism • No visa procedure for 72-hour ferry passengers in southern region	• Border formalities influence cross-border tourism attractiveness	• Border formalities influence cross-border tourism attractiveness • The e-visa practice has been delayed because of COVID-19 and technical issues	• Border formalities influence tourism attractiveness
	• Finland's economic support policies via EU development programs	• Russia's economic support policies via Russian federal practices	• In general, how important municipalities consider support and development of borderland SMEs		• The role of local and regional business development administrations is important for supporting, business advising, and consulting in borderlands
		• Russia has improved and asphalted the Karelian Road network since the 2010s	• Underdeveloped regional infrastructure • No regional horizontal transportation	• Underdeveloped regional infrastructure	• Functioning infrastructure conditions and existence of roads and other transportation facilities are important for tourism business

(*continued*)

Table 10.1 (continued)

Factor	Advantages		Disadvantages		Summary
Economic	• The tourism industry is generally growing				• The tourism industry is generally growing
	• Borderland shopping				• Possibilities for larger borderland shopping
				• Fluctuation of ruble makes difficult pricing of tourism services for international tour operators	• Fluctuation of ruble difficulties in pricing of tourism services
	• Great subsidiary business potential as a side-line business for local people				• Possibilities for subsidiary business potential to get extra income
	• SMEs already base their business partly on sustainability				• "Natural" regions' sustainability supports the interest of authentic Arctic regions
Sociocultural	• 20% of Finnish youth would like to move to the countryside • Rural multilocation and remote work interest		• Decreasing population • Net emigration; young people shifting to urban centers • Weak awareness and availability of rural services • Lack of professional workforce	• Decreasing population	• There exists an interest in moving to SPAs, but there is also a need for more information of rural services and workforce education to lure skilled persons
	• Networking is familiar to SMEs in tourism product development			• Networking is not familiar to SMEs in tourism product development	• Nationally different networking habits for operation

	• Changes in Western people's lifestyle, attitudes change toward work and leisure time, appreciation of sustainable living • Tourism possibilities based on people's livelihoods, languages, religions, values, family systems, and everyday manners of living (local traditional Karelian food and music culture)	• Karelian food and music culture		• Cross-border regions offer multiple tourism opportunities for lifestyle changes and work-leisure time and sustainable living for people • Appreciation of Karelian food and music offer business opportunities for cross-border SMEs
	• Increased independent domestic travel to avoid contact with many people because of COVID-19		• Increased domestic tourism partly making up for lost international tourism	• COVID-19 increased domestic tourism

(*continued*)

Table 10.1 (continued)

Factor	Advantages	Disadvantages	Summary
Technological	• Visibility, marketing, accessibility opportunities, 24/7 open marketplaces with digitalization	• Weak borderland digital networks • Lack of digital knowledge, enterprises' unwillingness to apply digital technology	• Opportunities of digitalization to increase tourism business
Environmental	• Less tourism, less environmental harms • Pure authentic nature and closeness to nature	• Away from mass tourism, less tourism for business	• Value of natural, unique, authentic environment increased
	• Increased general appreciation for green sceneries and natural and sustainable tourism	• Strong pressure to change (climate change, increased human operations, long-distance transport of air pollutants)	• Interest in pristine natural environments, but a need for land use planning
	• Four-season tourism as one of the major environmental business opportunities	• Climate change (warmer and sunnier summers, no-snow winters)	• Spring and autumn seasons unused
Legal	• Natural protection of international environmental agreements	• International agreements, international outlining, and different sanctions	• agreements protect nature and secure living conditions
	• COVID-19 instructions increased domestic tourism	• COVID-19 instructions decreased international tourism	• Consequences of COVID-19 restrictions
		• Sanctions by Western nations because of Russia's attack on Ukraine	• Cross-border collaboration and activities suspended

passports and are used to free cross-border movement (European Parliament, 2021) in Scandinavia and Europe, while Russians are used to free movement beyond their federal borders (Russian Federation, 1993). Free moving or border formalities are important factors that influence tourism attractiveness. Specifically, the most limiting factor for international tourism is the requirement of a visa to enter Russia. This hampers, for example, Asian tourists, who would like to quickly visit Russian Karelia while traveling to Finland. In some cases, it is possible to enter the Russian Federation without a visa, but it is only permitted for those traveling by ferry in the southern and southeast parts of Finland as part of an organized excursion group for a period of 72 hours (St. Peterline, 2020).

As an opportunity to increase tourism, however, temporary and short-stay visa issuance may open a totally novel cross-border tourism corridor to Russia. The Russian State Duma, the senate, and president adopted a bill on electronic visas in 2020. However, because of the COVID-19 pandemic and some technical issues, the implementation of the e-visa has been delayed. These e-visas will be able to be issued to 53 countries, including all Schengen Area country nationals, which will facilitate cross-border tourism.

Another political factor involves nations' economic support policies (or lack thereof) for tourism enterprises in SPAs. Finland supports tourism SMEs in accordance with the EU's general SME development programs (EC, 2020), and Russia does the same under the guidance of Russian federal practices (OECD, 2020b). The main issue, however, is how keen the states are to help cross-border SMEs in starting and/or developing their tourism enterprises. Sparsely populated borderlands may suffer from being peripheric locations because they are not the center of political interests. These regions would gain from economic support, business advising and consulting, and regional administrations. With this kind of support, these regions have better possibilities to keep the regions populated, which can create economic and political stability.

Finally, another important political obstacle that needs to be overcome to encourage the growth of the tourism industry in this region is the underdeveloped regional infrastructure. This issue concerns both railway and road travel. The Northern Ostrobothnia, Kainuu, and Russian Karelia regions are located parallel to each other. When traveling cross-border from one region to another, it is currently only possible to travel from east to west, or vice versa, by car because there are no public modes of transportation, such as buses or passenger trains, between these regions. Although Russia has improved the Karelian Road network since the 2010s (Saano, 2020), international cross-border tourism demands functional infrastructure conditions, especially public transportation. It is important to note that the lack of public transportation is not only a political issue, but also has remarkable sustainable implications. Private transport is one of the world's biggest sources of greenhouse gases, and reaching these sparsely populated regions requires long-distance traveling. To enter these destinations requires profound consideration of the means of travel. To develop a sustainable cross-border tourism industry, there is a need for investments to make these regions reachable by public transportation to minimize emissions.

Economic Factors

Globally, the pre-COVID-19 tourism industry grew 0.3% from 2015 to 2020 (IBIS World, 2021). In Finland, the number of international tourists has gradually grown since 2015, reaching 6.8 million overnight stays in 2018 (Visit Finland, 2021). In Russia, the number of inbound tourism visits has remained steady from 2016 to 2019, at about 24.4 million yearly tourism visits (Elagina, 2020). Travelers crossing the Finnish-Russian border were at their highest in 2013, at over 12.9 million. Prior to the COVID-19 pandemic—and because of the political situation leading to weakening of the ruble—there were around 9 million travelers (both ways counted) crossing the border annually (Makkonen et al., 2018). Overall, the number of tourists in Finland and Russia indicates great tourism potential and opportunities for the cross-border tourism industry, which could lead to an increase in the number of domestic and international visitors. Shopping tourism, which is popular in many Russian borderlands (see, e.g., Manakov et al., 2020; Sagan et al., 2018), including the regions in our case study, has greater possibilities and potential to grow on both sides of the border.

Currency is always an important issue in cross-border travel. As an economic factor, the currency fluctuation of the Russian ruble influences the possibility of subsidiary businesses in SPAs and related trends in the tourism industry. The fluctuation of any currency impacts pricing in the tourism industry, where pricing is often considered an impression of quality (Meidan, 1989). Hence, impressions of quality and currency fluctuations affect priced tourism services differently, making the pricing of tourism services and packages difficult for international tour operators. This is because the pricing period may be one or two years ahead, and if the currency swings constantly, international tour operators and their brokered customers may find other, easier, and more predictable priced tourism packages. Meanwhile, tourism has great potential as a side-line business that can broaden one's livelihood. People living in SPAs may earn extra money by renting summer cottages, offering adventure services, offering original national dishes to tourists, and so forth. The side-line business is usually periodical, based on Nordic seasons, and individuals can decide to what extent they will dedicate resources to the subsidiary business, depending on their primary source of income.

In tourism, short- and long-term trends guide customers' behaviors. With reference to current climate change conditions, a growing trend that increasingly informs human behavior is sustainability (Dolnicar, 2020; EC, 2020). Additionally, the COVID-19 period also changed people's behavior; preferences have shifted to closer locations, untouched nature, and less crowded places. Thus, sparsely populated, "authentic" areas can be perceived as suitable for sustainable tourism. In sparsely populated regions, the suitability effect is twofold. On the one hand, the means of transportation, food production, and usage of natural resources and energy production create significant emissions and costs because of long distances and not having a critical mass; therefore, transport costs per capita, for example, are relatively high, yet they constitute a basic premise for living in the regions. On the other hand, regions that offer untouched nature and a simple, more authentic experience of being in a remote area and silence bring people closer to nature and help foster understanding of the importance of environmental issues. Economically speaking, sustainability efforts support the interests in authentic tourism and protect its vulnerable environment at the same time. The typical activities offered by SMEs in these regions are naturally sustainable, based on locally produced food, home accommodation, nature trips, and sightseeing instead of building tourist attractions, which strengthens future tourism possibilities (i.e., Gavrila-Paven et al., 2015).

Sociocultural Factors

Demographic issues impact tourism development. The Northern Finnish region and the Eastern Finnish region both suffer from negative net emigration, as the number of emigrants exceeds the number of immigrants, shifting mainly young people away from rural areas to urban centers (Grunfelder et al., 2020). The population is also decreasing in the Republic of Karelia in Russia, where the number of inhabitants has decreased by around 6% during the years 2009–2020 (Federal Statistics Service, 2021). However, 20% of Finnish youth also express interest in settling down in the countryside and rural multilocation if there would be a possibility to earn a living there. Additionally, every third Finn is interested in remote work in the countryside and rural multilocation (Maaseutupolitiikka, 2020).

Based on previous studies (Maaseutupolitiikka, 2020) the most significant business obstacles in moving to rural regions are insufficient awareness and availability of countryside services and the lack of a professional workforce. From the point of view of entrepreneurship, to collaborate cross-border, it is essential to understand cultural behavior, local habits, and customs on both sides of the border. Finnish and Russian SMEs involved in the TourSME project were matched with each other in two workshops with the goal of producing cross-border tourism packages and networking. First, these SMEs created local multiday tourism packages that included different activities, for example, viewpoints, accommodation, and meals. Second, the SMEs introduced their local tourism package to the companies of the other country, with the aim of combining the outputs of local tourism packages into a common cross-border product. In this phase, cultural differences emerged. Finnish SMEs presented joint product packages, whereas Russian SMEs mainly presented company-specific products. There was therefore no collaboration between Russian SMEs to create broader tourism packages, nor were there discussions in the workshops which would encourage the Russian SMEs to create a joint tourism package. Makkonen et al. (2018) suggest that it is a question of very different business cultures; whereas Finnish companies tend to look for longer term business opportunities, Russian companies are accustomed to more immediate, short-term plans. This creates challenges as the tourism industry demands comprehensive services in cross-border tourism (accommodation, activities, meals), and therefore, networking in general is the base of any successful tourism business (Makkonen et al., 2018).

Sociocultural factors also support tourism activities in the region such as lifestyle, attitude changes towards work and leisure time, and migration to rural areas, which are becoming increasingly desirable places to live. SPAs offer an ideal environment for people who want to live surrounded by nature. Alongside this, attitudes toward work and leisure time are also changing, with work-life balance becoming highly valued. This has created great opportunities for SPAs to increase tourism by offering tourism packages according to the desires of people who are looking for ways to counterbalance work with life outside of work. Although SPAs suffer from the youth's rural-urban migration, tourism encourages migration back to these areas (Ala-Rämi, 2007) and entices young people to move back to rural regions.

The geographical and cultural circumstances of SPA tourism can offer many possibilities for different target groups, including the luxury and tranquility of being surrounded by nature in areas that are not densely populated, access to wilderness, authentic villages, and culturally lively small places. A glimpse into people's livelihoods, languages, religions, values, family systems, and everyday manners of living offer a huge amount of potential for these individuals to create tourism businesses. For example, although there may be commonalities in local food cultures across different countries, there are cross-border tourism opportunities to provide access to regional variances to tourists. In fact, both Finland and Russia share many traditional foods, such as Karelian stew and Karelian pastries, with various nuances based on common Karelian history and culture. Additionally, Karelian music culture is strongly appreciated on both sides of the border, offering more business opportunities for tourism SMEs.

Because of the recent COVID-19 pandemic, independent domestic travel has increased, partly replacing international tourism, hence favoring smaller places and SPAs. These regions possess the criterion that has been preferred by travelers during the pandemic: avoiding contact with many people. Thus, small-scale domestic accommodations with few people in less populated areas—and, therefore, with minimal physical contact—have become highly favored.

Technological Factors

New technologies are extremely important for the tourism industry (Hjalager, 2002), and technological factors are generally considered to increase competitiveness. This happens, for example, with new tourism product provision possibilities and modernization of servicing systems (Snieskiené, 2009). Governmental institutions stress that digitalization has a huge impact on all business endeavors. The usage of digitalization is important in customer acquisition on both sides of the border. Russians, especially, who have been investing strongly in well-being, health, and tourism, are the most active internet users (in Europe); over 110 million Russians search for information online every week.[4] In Finland, all tourism enterprises are guided to follow and update their digitalization readiness according to the Digital Roadmap (Visit Finland, 2019), stressing the importance of digital purchases and reservations for tourism products. Thus, digitalization offers great visibility and accessibility opportunities for tourism enterprises, as well as eases customer transactions.

Additionally, digital technology has increasingly allowed for remote work, allowing individuals to work from anywhere, which has led to greater mobility and allowed more individuals to freely choose their living areas without being limited by employment availability. More importantly, digitalization enables easy-to-access and cost-efficient marketing and visibility for tourism providers. Online marketplaces are also digital selling places, wherein potential customers may be introduced to producers' regions and encouraged to buy an arranged trip at the same time. For tourists, digitalization enables 24/7 open marketplaces and the possibility

[4] Kaupan liitto, Hannu Kyyhkynen, 29.05.2019, Uutiset, Venäjän talousnäkymät heikentyneet—digitaalinen kauppa kasvaa vauhdilla. https://kauppa.fi/uutishuone/2019/05/29/venajan-talousnakymat-heikentyneet-digitaalinen-kauppa-kasvaa-vauhdilla/.

of finding national and international tourism providers, as well as booking trips and accommodations online. Current challenges for increasing cross-border tourism are weak digital networks in border areas in both countries, a lack of digital knowledge, and enterprises' reluctance to rely on digital technology.

Environmental Factors

The impact that tourism has on the environment is huge. Climate change is recognized as having an impact on tourism and outdoor recreation, both directly and indirectly, because the changing natural environment (Nicholls, 2006) also influences the changes in nature-based tourism. In Northern Finland and Northern Russia, climate change has had a positive impact on summer tourism because of the longer, sunnier, and warmer seasons, as well as warmer natural waters. However, it can also have an impact on winter tourism because of potential problems regarding the lack of snow in the southern parts of Europe (Saarinen & Tervo, 2006). Thus, sustainability in tourism is the best way to reduce the environmental harm caused by tourists (Dolnicar, 2020).

All-season tourism is one of the major environmental business opportunities to be considered to grow the tourism industry in SPAs. In the northern parts of Europe, there are distinct opportunities for all four seasons. Snowy winters offer activities such as skiing and dog-sled racing, while northern summers with the midnight sun are also well-known tourism attractions. Light-green springs and colorful autumns offer berry-and mushroom-picking experiences for tourists coming from countries with less seasonal variations. Some SPAs are unaware of the attraction of the spring and autumn seasons and these seasons' activities could thus be marketed more by tourism businesses.

The vulnerability of the environment must be considered by local tourism industries. Specifically, studied areas such as SPAs are clear and natural and thus vulnerable. However, they are still less vulnerable than polar regions (see Hall & Saarinen, 2010). This region cannot be imagined as growing as a mass tourism destination, but pristine natural environments can be sustained, even with an increase in tourism; this can be done through careful sustainability planning which looks to balance tourism services and the region's natural state (Hakkarainen & Tuulentie, 2008; Saarinen, 2007). Although the region meets the same global challenges as all other tourism destinations, such as climate change, increased human operations, and the air pollution of long-distance transport, tourism in the regions can be operated by highlighting better community/regional planning and land use planning. This requires putting tourism and all its impacts in the perspective of larger regional socioeconomic contexts (Saarinen, 2007). In practice, this means that local actors and resources are at the center of tourism development and lead the development work (Hakkarainen & Tuulentie, 2008; Saarinen, 2007).

Local and regional tourism sectors may flourish by putting efforts into sustainable land and natural resource usage. However, globally, the influence of climate change means warmer and sunnier summers. Thus far, there has been no influence on the winters in the studied regions. From an environmental perspective, snowy winters attract more tourists than warm summers. Global climate warming negatively influences snow tourism because of declining snow coverage (Mingle, 2020).

Legal Factors

Legal factors, such as international agreements, international outlining, and different sanctions, are often considered to have a negative influence on tourism. However, as previously discussed, the electronic visa that will enable cross-border travel between the EU and Russian Federation will have a positive impact on the tourism industry in these regions. Overall, international agreements, such as the Nature Conservation Act,[5] and their associated sanctions guide people's and enterprises' operations to keep nature and the environment as unique, pure, and natural as possible, in turn securing the living conditions of indigenous people.

Ensuring that the regions maintain their unique charms also offers interesting tourism potentialities. Thus, international agreements and possible sanctions can ultimately protect the tourism business in these

[5] https://environment.ec.europa.eu/international/protecting-biodiversity-worldwide-towards-international-agreement-cop-15_en.

regions. The COVID-19 pandemic, with its international and national legal restrictions, has influenced international and domestic tourism in two ways. First, the spread of COVID-19 stopped international tourism because of international aircraft movement restrictions. In turn, increased domestic tourism partly filled the gap left by a lack of international tourists in several countries, thus creating new tourism opportunities. Individual well-being, tourism, good food, and a sense of leisure are now appreciated factors for individuals looking to travel (Markkinointi & Mainonta, 2020). These business opportunities are present in our study region's tourism industry as well. After our research was conducted, major new legal factors arose because Russia's attacks on Ukraine violated international laws, thereby leading to sanctions by Western nations, which means that all collaboration and cross-border activities have been stopped for the time being.

Discussion and Conclusion

This chapter has focused on tourism industries in cross-border regions, specifically Northern Ostrobotnia, Kainuu, and Russian Karelia. These regions and the nature and culture they offer for tourism are often unnoticed by larger audiences. The regions may not have the same reputation as the polar regions, but they can offer unique tourist experiences and "otherness" (Irvine & Anderson, 2007) that are economically and ecologically reachable in a safe environment. In this way, these regions possess enormous potential for developing rural tourism services.

The importance of tourism is significant in remote areas for the local people, but it also keeps the region vital and populated. Questions surrounding sustainability naturally arise when talking about the growth of the tourism industry and increased mobility. To return to the concept of sustainable tourism defined at the beginning of this chapter and the needs of the inhabitants of the study area and of the regions, if the challenges we have identified can be met, the study area will have much to offer to those interested in unique natural and cultural sites.

Seizing these opportunities will create better conditions for new actors to join the tourism sector, such as farmers, fishers, and well-being entrepreneurs, who can utilize business opportunities and offer part-time services for tourists and strengthen the possibility of earning livelihoods in their own regions via sustainable tourism. Digitalization, social entrepreneurship, and dynamic changes in working life can create solutions to social, cultural, or environmental issues, as well as diversify the motives and operating models of traditional, purely economy-driven business making. Instead of creating tourist services that attract mass tourism, less should be considered as more in many cases; "live like locals" experiences and multilocation work solutions are considerable and attractive factors in sparsely populated rural regions with rich natural offerings that are also sustainable activities.

Considering the increasing importance of sustainable travel, the studied regions are in an ideal position to use existing local strengths to create tourism services and ensure that the principles of sustainable tourism are preserved (Hunt, 2018) because there is also growing interest in individual, unique experiences. The regions possess plenty of tourism opportunities that place nature at their core, including the ability to offer simple experiences and relaxing atmospheres in SPAs, which may influence business opportunities that have the potential to change net emigration (Grunfelder et al., 2020). The most important way to make tourism more sustainable is to change people's behaviors and to put more considerable effort in creating unique experiences and services for tourists, which can support sustainable behaviors (Dolnicar, 2020). Furthermore, experiences based on peaceful and calm environments, fresh air, and untouched nature are becoming increasingly valued by people who experience stressful situations in their daily lives. Economically, developing sustainable and nature-based activities (Dolnicar, 2020) strengthens the attractiveness of the tourism site and enhances environmental factors while protecting the regions' authentic environment.

Based on discussions among the SMEs during the TourSME project activities, the main problem identified was that tourism SMEs possessed a low capability for offering border-crossing services to international tourists, partly because of language, but also because of a lack of awareness of the needs of tourists. Despite some region-specific challenges, many of the business challenges are shared on both sides of these cross-border regions, and a strong, shared understanding exists that cross-border tourism SMEs will enhance the economic competitiveness of these regions.

In sparsely populated rural regions, existing human resources are quite limited, and services are rather scant. Although these regions are small, they still have business environments with broad opportunities. As mentioned before, the current research was carried out prior to the invasion of Ukraine. The war in Ukraine has changed many things, among others also our views on the possibilities and challenges that cross-border tourism in this region possesses. The border that has been closed since March 2020 will remain closed for an indeterminate period of time. Yet the present research, as well as the ongoing situation in Ukraine, has given us new perspectives on the fragility of such regions and their associated tourism opportunities and activities.

References

Ala-Rämi, K. (2007). Communication and distance in collaboration between high-technology enterprises in Northern Finland. *European Planning Studies, 15*(8), 1047–1062. https://doi.org/10.1080/09654310701448212

Carbone, G. (2018, December 11). *Expert analysis on geographical specialties: Mountains, Island and Sparsely Populated Areas. Cohesion policy 2014–2020.* https://ec.europa.eu/regional_policy/sources/docgener/studies/pdf/expert_analysis_geographical_specificities_en.pdf

CIPD Community. (2021, January 30). *PESTLE analysis.* https://www.cipd.co.uk/knowledge/strategy/organisational-development/pestle-analysis-factsheet#gref

Dolnicar, S. (2020). Designing for more environmentally friendly tourism. *Annals of Tourism Research, 84*, 102933. https://doi.org/10.1016/j.annals.2020.102933

EC, European Commission. (2020, October 5). *Doing business in the EU, Finance and support.* https://ec.europa.eu/info/business-economy-euro/doing-business-eu/finance-and-support_en

Elagina, D. (2020, September 14). *Number of inbound tourist arrivals in Russia 2014–2019.* https://www.statista.com/statistics/888119/russia-international-tourist-arrivals/#statisticContainer

European Parliament. (2021, January 22). *Free movement of persons*. https://www.europarl.europa.eu/factsheets/en/sheet/147/henkiloiden-vapaa-liikkuvuus
Federal Statistics Service. (2021, January 22). *Population: NW: Republic of Karelia, view Russia's population: NW: Republic of Karelia from 1989 to 2020 in the chart*. https://www.ceicdata.com/en/russia/population-by-region/population-nw-republic-of-karelia
Gavrila-Paven, I., Constantin, B., & Dogaru, L.-D. (2015). Advantages and limits for tourism development in rural areas (case study Ampoi and Mureş Valleys). *Procedia Economics and Finance, 32*, 1050–1059. https://doi.org/10.1016/S2212-5671(15)01567-1
Gregorić, M. (2014). *Pestel analysis of tourism destinations in the perspective of business tourism (MICE)*. Tourism and Hospitality Industry 2014, Congress Proceedings Trends in Tourism and Hospitality Industry. https://www.free-management-ebooks.com/faqst/pestle-09.htm#ixzz6s6GBMDZK
Grunfelder, J., Norlén, G., Randall, L., & Sánchez, G. (Eds.). (2020). *State of the Nordic region 2020*. https://doi.org/10.6027/NO2020-001
Hakkarainen, M., & Tuulentie, S. (2008). Tourism's role in rural development of Finnish Lapland: Interpreting national and regional strategy documents. *Fennia, 186*(1), 3–13.
Hall, C. M., & Saarinen, J. (2010). Polar tourism: Definitions and dimensions. *Scandinavian Journal of Hospitality and Tourism, 10*(4), 448–467. https://doi.org/10.1080/15022250.2010.521686
Hjalager, A. M. (2002). Repairing innovation defectiveness in tourism. *Tourism Management, 23*(5), 465–474. https://doi.org/10.1016/S0261-5177(02)00013-4
Hunt, E. (2018). Residents in tourism hotspots have had enough. So what's the answer? *The Guardian*. https://www.theguardian.com/cities/2018/jul/17/residents-in-tourismhotspots-have-had-enough-so-whats-the-answer
IBIS World. (2021, July 9). *Report of global tourism industry trends (2015–2020)*. https://www.ibisworld.com/global/market-research-reports/global-tourism-industry/
Irvine, W., & Anderson, A. R. (2007). The vulnerability of peripheral tourism: The rapid disenchantment of peripheral attraction. In D. K. Müller & B. Jansson (Eds.), *Tourism in peripheries: Perspectives from the far north and south* (pp. 109–123). CABI. https://doi.org/10.1079/9781845931773.0109
Izotov, A., & Laine, J. (2013). Constructing (Un)familiarity: Role of tourism in identity and region building at the Finnish–Russian border. *European*

Planning Studies, 21(1), 93–111. https://doi.org/10.1080/09654313.2012.716241

Karelia CBC. (2021, July 22). Karelia cross-border operation. https://www.kareliacbc.fi/index.php/en; https://www.kareliacbc.fi/index.php/en

Maaseutupolitiikka. (2020, October 10). *Maaseutubarometri. (2020) Osa 2: Yrittäjyys ja elinkeinot maaseudulla.* https://www.maaseutupolitiikka.fi/uploads/MANE-raportit/Maaseutubarometri-2020-tutkimusraportti-osa-2-FINAL.pdf

Maher, P. T., Gelter, H., Hillmer-Pegram, K., Hovgaard, G., Hull, J., Jóhannesson, G. T., & Pashkevich, A. (2014). Arctic tourism: Realities and possibilities. In L. Heininen (Ed.), *Arctic yearbook 2014: Scholarly articles, section II: Regional economy and property* (pp. 290–306) www.arcticyearbook.com

Makkonen, T., Williams, A. M., Weidenfeld, A., & Kaisto, V. (2018). Cross-border knowledge transfer and innovation in the European neighbourhood: Tourism cooperation at the Finnish-Russian border. *Tourism Management, 68*, 140–151. https://doi.org/10.1016/j.tourman.2018.03.008

Manakov, A. G., Kondrateva, S. V., & Terenina, N. K. (2020). Development of cross-border tourist and recreational regions on the Karelian section of the Russian-Finnish border. *Baltic Region, 12*(3), 140–152. https://doi.org/10.5922/2079-8555-2020-2-9

Markkinointi & Mainonta. (2020, September 14). *Magazine for marketing communication and media professionals.* https://www.marmai.fi/uutiset/tassa-ovat-suomalaisten-korona-ajan-matkailutrendit-yksi-kotimaan-kohde-haastaa-helsingin-tosissaan/952fbd2a-fc91-4d14-b880-2408bc89191e

Meidan, A. (1989). Pricing in tourism. In S. F. Witt & L. Moutinho (Eds.), *Tourism marketing and management handbook* (pp. 305–309). Prentice Hall.

Mingle, J. (2020). IPCC special report on the ocean and cryosphere in a changing climate. *The New York Review of Books, 67*(2020), 49–51.

Narvik University College. (2010). *Arctic regions: Concerns, threats and potential challenges.* Project report for PRETEAR.

Nicholls, S. (2006). Climate change, tourism and outdoor recreation in Europe. *Managing Leisure, 11*(3), 151–163. https://doi.org/10.1080/13606710600715226

OECD. (2020a, October 12). *OECD tourism trends and policies 2020.* OECD Publishing. https://doi.org/10.1787/6b47b985-en

OECD. (2020b). OECD iLibrary. *Financing SMEs and enterprises 2020: An OECD scorecard. 38. Russian Federation.* https://www.oecd-ilibrary.org/

sites/4034a9a8-en/index.html?itemId=/content/component/4034a9a8-en
POLITICO. (2022). Finland to close border to Russian tourists. https://www.politico.eu/article/finland-close-border-frussian-tourists/
Prokkola, E.-K. (2007). Cross-border regionalization and tourism development at the Swedish-Finnish border: "Destination Arctic Circle". *Scandinavian Journal of Hospitality and Tourism, 7*(2), 120–138. https://doi.org/10.1080/15022250701226022
Ragulina, Y. (2020, August 31). Personal communication.
Russian Federation. (1993, November 20). *CIS legislation. The Russian Federation Law No. 5242-1*. https://cis-legislation.com/document.fwx?rgn=1446
Saano, V. (2020). *Runokyläkierros Vienan Karjalassa*. https://www.venajanaika.fi/3-2020/runokylakierros-vienan-karjalassa/
Saarinen, J. (2007). Tourism in peripheries: The role of tourism in regional development in northern Finland. In D. K. Müller & B. Jansson (Eds.), *Tourism in peripheries: Perspectives from the far north and south* (pp. 41–53). CABI. https://doi.org/10.1079/9781845931773.0041
Saarinen, J., & Tervo, K. (2006). Perceptions and adaptation strategies of the tourism industry to climate change: The case of Finnish nature-based tourism entrepreneurs. *International Journal of Innovation and Sustainable Development, 1*(3), 214. https://doi.org/10.1504/ijisd.2006.012423
Sagan, I., Kolosov, V., Studzińska, D., Zotova, M., Sebentsov, A., & Nowicka, K. (2018). The local border traffic zone experiment as an instrument of cross-border integration: The case of Polish-Russian borderland. *Geographia Polonica, 91*(1), 95–112. https://doi.org/10.7163/GPol0102
Snieskiené, G. (2009). Isorinés aplinkos veiksniu poveikio eksporto kainodaros strategijai vertinimas. *Economics & Management, 14*, 968–975.
St. Peterline. (2020, December 10). *Visa free rules: RF Government Resolution №397*. https://stpeterline.com/en/-/visa-free-rules
Statistics Finland. (2022). *Statistics Finland, population*. https://www.stat.fi/til/vrm_en.html
Stoffelen, A., & Vanneste, D. (2017). Tourism and cross-border regional development: Insights in European contexts. *European Planning Studies, 25*(6), 1013–1033. https://doi.org/10.1080/09654313.2017.1291585
Taipale-Erävala, K., Heilmann, P., & Lampela, H. (2014). SME competence transformation: A case study on industrial structural change. *International Journal of Business Innovation and Research, 8*(3), 265–281. https://doi.org/10.1504/IJBIR.2014.060826

Timothy, D. J. (1999). Cross-border partnership in tourism resource management: International parks along the US-Canada border. *Journal of Sustainable Tourism, 7*(3-4), 182–205. https://doi.org/10.1080/09669589908667336

Ulkopolitiikka. (2019, March 7). *Kajoaako Putin perustuslakiin?* Foreign policy. https://ulkopolitiikka.fi/lehti/1-2019/kajoaako-putin-perustuslakiin/

United Nations. (2021, October 8). Department of economic and social affairs, population dynamics. *Total population – Both sexes*. https://population.un.org/wpp/Download/Standard/Population/

United Nations. (2022, August 12). *Sustainable lifestyles*. https://www.unep.org/explore-topics/resource-efficiency/what-we-do/sustainable-lifestyles

UNWTO. (2022, October 2). UNWTO tourism market intelligence and competitiveness. Impact of the Russian offensive in Ukraine on international tourism. https://www.unwto.org/impact-russian-offensive-in-ukraine-on-tourism

Visit Finland. (2019, November 1). *Digital map of Finnish tourism, 1st update November 2019*. https://www.businessfinland.fi/492e08/globalassets/julkaisut/visit-finland/tutkimukset/2019/suomen_matkailun_digitiekartta_2019.pdf

Visit Finland. (2021, December 12). *Rudolf statistics. Population*. http://visitfinland.stat.fi/PXWeb/pxweb/fi/VisitFinland

Webster, C., & Ivanov, S. (2013). *Tourism as a force for political stability. The international handbook on tourism and peace*. https://ssrn.com/abstract=2311176

Yurievna, M. F. (2011). Tourism development in the Republic of Karelia: Problems and perspectives. In *Proceedings of the International Conference on Tourism & Management Studies* (Vol. I, pp. 60–69). ALGARVE.

11

Environmental and Tourism Policies for the Decarbonization of the Tourism Sector

Yoshihiro Hamaguchi

Introduction

Tourism is a huge industry, accounting for approximately 10% of the world's GDP and 7% of its exports. Its development can lead to increased incomes, employment, tax revenues, and the earning of foreign currency; governments are thus implementing aggressive growth strategies for tourism. The tourism industry is heavily dependent on the natural environment, but the relationship between tourism and the environment is complicated. Buckley (2011), who typifies previous studies on tourism and the environment, points to the need for future research on tourism's effects on climate change and its potential irreversible environmental damage. The main effects of climate change on touristic destinations are increased temperature, decreased snowfall, more frequent storms, both decreased and increased precipitation, a rise of sea level, increased sea

Y. Hamaguchi (✉)
Department of Economics, Faculty of Economics and Business Administration, Kyoto University of Advanced Science, Kyoto, Japan
e-mail: hamaguchi@kyoto-econ.ac.jp; prestidigtateur@gmail.com

T. Walker et al. (eds.), *Sustainable Tourism*,
https://doi.org/10.1007/978-3-031-43528-7_11

surface temperature, changes in marine ecosystems, more frequent forest fires, and changes in soil (UNWTO & UNEP, 2008). Though an increase in sea temperature may encourage tourism as it will allow for more extended tourist seasons in some beach areas, it also increases precipitations due to water vapor and causes flooding. Particularly problematic is the fact that tourism increases greenhouse gas (GHG) emissions as it increases traffic. Thus, the tourism industry is not only affected by climate change but also causes climate change in turn.

Among the modes of transport for international tourism such as air, rail, car, and coach-aviation, air has the highest impact on GHG emissions. The United Nations World Tourism Organization (UNWTO) and the United Nations Environment Programme (UNEP) estimate the emissions of each mode of transport in the European Union using the carbon dioxide factor (kg/pkm); the estimates of global average are 0.1333 for cars, 0.129 for aviation, 0.027 for rail, and 0.022 for horse-drawn carriages (UNWTO & UNEP, 2008). When broken down by distance, an estimate of 0.206 for short-haul air travel of <500 km has the highest carbon dioxide emissions and 0.154 for 500–1500 km almost exceeds the 0.1333 of cars. Of the estimated 26400Mt of global carbon dioxide emissions in 2005, the tourism sector is estimated to have accounted for 1302Mt (4.940%). Air transport accounted for the largest share at 40%, 32% for cars, and 21% for accommodation (UNWTO & UNEP, 2008).

In response to this data, the UNWTO (2015) aims to achieve sustainable tourism under the United Nation's Sustainable Development Goals (SDGs). The current tourism industry is still dependent on wasteful resource use and on unsustainable consumption and production. Therefore, the key metrics to consider here are SDGs 8, defined as Decent Work and Economic Growth, SDGs 12, defined as Responsible Consumption and Production, and SDGs 17, defined as Partnerships for the Goals. These measures are aimed to be achieved by 2030. The tourism industry is also highly dependent on global economic and security trends. This fact is also recognized by SDGs 11, 14, 15, and 16. SDG 13 also points to natural disasters and the loss of biodiversity through climate change and identifies these as important issues for the tourism industry in the future. Lenzen et al. (2018) predict that the tourism industry, expected to grow by an average of about 4% per year globally, will become a carbon-intensive industry and a major global emitter of GHGs in the future. They found that from 2009 to 2013, as tourist spending increased from \$2.5 to \$4.7 trillion, the industry's carbon footprint rose rapidly from 3.9 GtCO2e to 4.5 GtCO2e, with the industry accounting for approximately 8% of global GHGs. This outcome was possible because of how slowly governments introduced carbon taxes and emissions trading on air transport for tourism. The Convention on International Civil Aviation Organisation (Chicago Convention) has a "non-discrimination principle" prohibiting discriminatory treatment between parties, while the Kyoto Protocol has a "common but differentiated principle" exempting developing countries from GHG reduction obligations. Such inconsistencies have prevented the introduction of environmental regulations in the aviation industry. However, in 2016, to address climate change, EU member states and ICAO member states agreed on a global market-based measure, to impose the EU-Emissions Trading System on all airlines flying within the EU in 2020 (European Commission, 2017).

Evidently, environmental regulations must be introduced to the tourism industry, especially for air transport. However, this will not necessarily lead to more sustainable tourism because strict environmental regulations would hinder the development of the tourism industry. In UNWTO (2015), policymakers in developing countries believed that sustainable tourism would accelerate the SDGs through multiplier effects with other industries. While competitiveness and profitability remain key drivers of sustainability, the tourism industry, which emits more than 5% of global greenhouse gases, will need to reduce its emissions below current levels. Therefore, it is necessary to clarify how environmental policies affect the relationship between tourism, economic growth, and the environment. In this chapter, we examine representative empirical and theoretical analyses, present mechanisms of this relationship, and recommend polices.

Tourism-led Growth Hypothesis

Empirical Facts

Before discussing sustainable tourism, we will outline the tourism-led growth hypothesis (TLGH), because the impact of tourism on economic growth is under debate. The TLGH, developed by Balagner and Cantavela-Jordá (2002) is the first empirical demonstration of the fact that tourism stimulates economic growth. Brida et al. (2016) found that TLGH is valid, although many studies have focused on the tourism multiplier effect, where the tourism sector's expansion leads to GDP expansion through a multiplier effect. However, these studies were subject to sample selection bias, as they largely focused on countries with a developed tourism industry, and few African and Middle Eastern countries were analyzed.

The TLGH has been categorized as (i) a unidirectional relationship between tourism development and economic growth (TLGH), (ii) a bidirectional relationship, (iii) an inverse relationship between economic growth and the development of tourism (economy-driven tourism development; EDTG), (iv) a neutral relationship, and (v) growth hindered by tourism development. Pulido-Fernández and Cárdenas-García (2021), using a sample of 143 countries from 1993 to 2017, point out factors that determine the bidirectional relationship between tourism growth and economic development. Through a meta-regression analysis of 545 estimates from 113 studies, Nunkoo et al. (2020) conclude that TLGH has a statistically significant relationship. However, they mention the possibility of publication bias and state that this estimate is highly sensitive to country data, specifications, estimation characteristics, and time horizons.

In small island developing states, dependence on tourism may hinder economic development because it inhibits the growth of other industries. Cárdenas-García et al. (2015) define economic development in terms of human development indicators, life expectancy at birth, infant mortality in less than one year, adult literacy, and public expenditure on health and education. Tourism growth leads to economic development in developed countries, but hinders economic development in developing countries. Touristic countries rely on natural/environmental tourism resources. However, a "resource curse" may occur in these countries, in which abundant natural resources inhibit economic development. Abundant tourism resources crowd out investment in human capital and the tourism boom crowds out industrial production (Deng et al., 2014). The degree of democratization may also affect the effectiveness of the TLGH. Antonakakis et al. (2019) find that the TLGH is established in developing countries with undemocratic bureaucracies and little tourism specialization, while efficient democracies establish a two-way relationship. Das and Dirienzo (2010) state that corruption reduction improves tourism competitiveness between countries and that its marginal effect is greater in developing countries than in developed countries. The impact of tourism growth on economic development varies according to the stage of tourism development (Chiu & Yeh, 2017). Lee and Chang (2008) find that the growth-promoting effect of tourism is greater in non-OECD countries than in OECD countries and that it is greatest in sub-Saharan African countries. Hence, democratic movements and political reforms in developing countries indirectly support TLGH effects. Efficient administrative systems can help prevent rent-seeking from environmental tourism resources and discourage the resource curse. However, Croes et al. (2018) argue that in Malta, the relationship between tourism specialization and economic growth is not self-evident and there is a negative correlation between investment and quality of life. Song and Wu (2022) argue that the theoretical background of TLGH is not statistically significant, as the mechanisms by which tourism development is linked to economic growth, such as how tourism specialization associated with tourism exports improves economic productivity, are unclear. Where there is a theoretical mechanism for a causal relationship between two variables, the Grainger's causality test indicates that relationship. Therefore, theoretical analysis is required to reveal the mechanism.

Theoretical Mechanisms

Recent studies on the TLGH have relied on the endogenous growth theory; Pina and Martínez-García (2013), for instance, consider how readily available information about a touristic destination can increase its popularity. Tourists refer to information in guidebooks and on the internet when deciding which destinations to visit and are more likely to visit locations about which there is abundant information. An increase in tourism in a location leads to it becoming increasingly popular among tourists, which in turn increases a destination's economic growth. This result is empirically supported by Lee et al. (2019), who specify the bias against halo effects of perceived destinations. Alternatively, tourism information regarding a lack of touristic resources, such as historical sites, hinders economic development. These touristic destinations could attract tourists by building tourist facilities, however other industrial sectors crowd out investments (Tugcu, 2014). Using a Bayesian dynamic stochastic general equilibrium model, Liu and Wu (2019) identify an externality effect between the tourism sector and other industries through public services, in addition to physical and human capital. An improvement in productivity in one sector improves productivity in another sector, which in turn leads to an expansion of output and higher economic growth. The simulation analysis for Spain shows that when the productivity of the economy (tourism sector) improves, inbound demand (domestic tourism consumption) increases beyond the demands of domestic tourism (inbound consumption).

Building on Hjalager (2010), who surveys the relationship between international tourism and innovation, Albaladejo and Martínez-García (2015) analyze the impact of innovation on tourism development using a lab-equipment model. Butler's (1980) tourism area life cycle (TALC) hypothesis discussed below states that a stagnant tourism destination moves into a period of maintenance and declines because its carrying capacity cannot keep up with the increase in tourists. They believe that when innovation improves the carrying capacity of a destination, it welcomes more tourists and moves into a regenerative phase. When an innovation increases the variety of intermediate goods that are raw materials for final goods, it expands the range of tourism and the final goods required for investment. Greater investment in tourism will lead to improved carrying capacities. This effect increases the number of tourists and the income they bring with them. In another study on tourism and innovation that relies on endogenous economic growth theory, Barrera and Garrido (2018) find an inverse U-shaped relationship between public holidays and economic growth rates. An increase in public holidays increases demand for tourism goods, thereby encouraging innovation and increasing the economic growth rate. Innovation increases the variety of intermediate goods that can be used as raw materials for final goods and consequently increases the production of tourism goods. Alternatively, an increase in public holidays reduces researchers' working hours, which stagnates innovation and thus reduces the rate of economic growth, per Parrilla et al. (2007) and Brida et al. (2016)'s earlier theoretical models. The interdependent relationship between tourism and growth has been developed through the feedback of both theoretical research and empirical analysis. By adding the environment to this relationship, we can understand the mechanisms of sustainable tourism.

Sustainable Tourism

Empirical Facts

The relationship among tourism, economic growth, and the effects of tourism on the environment is under debate. Cerina et al. (2010) analyze sustainable tourism through an economic lens, though no clear definition of sustainable tourism has yet been established. Liu (2003) argues that there are many false assumptions and arguments against this concept and critically examines it in terms of the role of demand in tourism, the nature of tourism resources, the improvement of intra-generational equilibrium, the role of tourism in promoting socio-cultural progress, the measures of sustainability, and the forms of sustainable development. Consequently, a systematic outlook and interdisciplinary approach are essential for deriving its scientific definition. Johnston and Tyrrell (2005), through their optimal control model of tourism and the environment, attempt to clarify what should be considered sustainable tourism, for what purpose, and to what level. They conclude that there is no universally optimal level of sustainability. Besides, the situation of unsustainability for either tourism companies or the local population is caused not only by policies that cause excessive environmental destruction but also policies that over-maintain the environment. In order to understand what factors are involved in sustainable tourism and what their interrelationships are, we will review, in this section, empirical facts and related mechanisms.

Experts are divided as to what conclusions should be drawn from empirical findings on sustainable tourism. They cannot agree that the development of tourism reduces GHG while it promotes economic growth. For example, Lee and Brahmasrene (2013)'s study, which analyzes EU member states, notes that an increase in tourism receipts of 0.1498% led to an increase in economic growth of 0.498% and a decrease in carbon dioxide emissions of 0.105%. Ahmad and Ma (2022) show that a 1% development in tourism leads to a 0.3862% reduction in carbon dioxide emissions in Asia. Paramati et al.'s 2017 study observing 26 developed and 18 developing countries finds that tourism development increases economic growth rates in both developed and developing countries. These facts led to TLGH and the observation of an ensuing environmental Kuznets curve between tourism and carbon dioxide emissions. At low GDP per capita levels, the tourism industry stimulates economic development but emits large amounts of carbon dioxide. However, when a high GDP per capita level is reached through further economic development, citizens become more environmentally conscious and energy-efficient innovations will occur. As a result, the carbon dioxide emissions of the tourism industry decrease. Thus, the tourism industry transforms from carbon-intensive to non-intensive. In contrast, Zaman et al.'s (2016) study of 34 countries found an environmental Kuznets curve, wherein tourism development and energy demand increased carbon dioxide emissions, which led to EDTG. They argue that there is a trade-off between tourism and the environment. Wang and Wang (2018), covering OECD countries, find that tourism growth leads to more carbon emissions and hampers tourism development. Zhang and Zhang (2021) find that China has a long-run bidirectional relationship between carbon dioxide emissions and GDP, carbon dioxide emissions and tourism, and GDP and tourism. Papavasileiou and Tzouvanas (2021) call the inverted U-shaped relationship between the amount of economic and carbon performance among tourism enterprises the tourism carbon Kuznets curve hypothesis. With certain reservations, they conclude that these results are robust according to various specifications. Thus, there is a need for further research on the empirical relationship between tourism, economic growth, and the environment. A theoretical analysis of this relationship would provide useful insights for future research.

Economic Growth Model

Early studies analyzed sustainable tourism through exogenous growth models. Lozano Ibáñez et al. (2004) point to an externality effect where environmental quality improves the productivity of the tourism goods sector. In a dynamic model consisting of an agricultural goods sector and a tourism goods sector, they consider that exogenous economic growth rates increase the relative price of tourism goods and find that it is socially desirable to halt the expansion of the tourism goods sector. The increase in the price of tourism goods owing to economic growth has led to an expansion in the production of tourism goods. This expansion has reduced the production of agricultural goods, which in turn has expanded available land and worsened the quality of the environment. The deterioration of the environment reduces the productivity of the tourism goods sector through externality effects, and the social optimum of the equilibrium is undermined. This is in contrast to the empirical results found by Romão (2020) for a sample of Western countries. His results show that tourism specialization, where tourism and agriculture are integrated, positively affects both economic growth and resilience. Moreover, Tugcu (2014) finds that growth through tourism is more likely to occur in the Mediterranean region of Europe. Thanks to inter-industry externality effects in this area, development of the tourism industry encourages development of other industries, which in turn promotes economic growth. The growth-promoting effects of tourism may vary depending on whether environmental variables are considered.

Cerina (2007) concludes that pollution abatement activities through tourism tax revenues determine sustainable tourism. A larger number of tourists brings in higher revenue and leads to higher consumption in any given destination. However, this increase also negatively affects the environment of that destination due to an increase in congestion. This in turn worsens the productivity of tourism goods. Taxation of the tourism sector reduces the demand for tourism goods but improves environmental quality by reducing tourists, as well as the input of public pollution abatement goods through tax revenues. This improves the productivity of the tourism goods sector through externality effects and consequently increases the production of tourism goods. According to Gómez et al. (2008), an increase in tourists affects congestion, the quality of the environment, and public goods. A decrease in tourists due to an increase in the accommodation tax improves environmental quality but worsens economic welfare due to lower consumption of tourism goods. Alternatively, this decrease reduces the congestion of public goods, improves the productivity of accommodation through public goods, and increases the level of consumption of tourism goods. Hence, the accommodation tax can have a double dividend: improved environmental quality and improved welfare. Adedoyin et al. (2021) find that a 10% increase in tourism taxes leads to a 5.4% decrease in tourism demand in the Maldives. They argue that fiscal policy efficiency is important for increasing inbound demand. These results are consistent with the theoretical results.

Recent studies have analyzed these findings in an endogenous growth model, where the improved environmental quality improves productivity and, consequently, economic growth. These analyses consider the pollution abatement activities of firms and identify a win-win relationship between tourism and the environment. Marsiglio (2015) finds that environmental quality influences the number of tourists, but as tourism increases, environmental quality deteriorates (congestion effect). Deteriorated environmental quality is improved by the pollution abatement activities of tourism firms. The tourism goods produced are used for consumption by tourists, investment in tourism equipment, and pollution abatement activities. In equilibrium, improvements in environmental quality lead to economic growth when tourists strongly prefer the environment. Ouattara et al. (2019) also find a win-win relationship through an in-depth analysis of transitional dynamics, which suggests that the promotion of ecotourism leads to sustainable tourism. Focusing on the spillover effects of over-tourism on sustainability in the economic and social spheres rather than its negative effects on the environment, Schubert and Schamel (2021) consider tourism quantity, such as the number of tourists, and tourism quality, such as the number of hotel rooms. They argue that when there are too many tourists, the marginal utility of tourism consumption is reduced due to congestion in the tourist area. Therefore, in addition to limiting the number of tourists, increasing production through tourism investment is necessary to maintaining sustainability. If environmental quality is considered in their model, in addition to reducing the congestion effect of limiting the number of tourists, the removal of pollution through tourism investment will improve environmental quality, resulting in a win-win relationship between tourism and the environment.

Tourism Area Life Cycle Model

The tourism industry often undergoes structural changes, and studies have emerged that look at the relationship between tourism, the environment, and growth from the perspective of structural change using Butler's (1980) Tourism Area Life Cycle Hypothesis (TALC). In the TALC, which uses a geographic approach, the development stage of a tourism area is defined by the number of tourists, and the tourism area follows an S-curve development stage as follows: exploration, involvement, development, consolidation, stagnation, and rejuvenation or decline. Using this hypothesis to empirically investigate the structural change in tourism in Chinese provinces, Zuo and Huang (2018) find that the inverse U-shaped relationship between tourism development and economic growth becomes N-shaped owing to the improvement of profitability and economic growth through capital investment. In the early stages of development, specialization in the tourism industry attracts tourists and the economy grows rapidly through increased tourism income. However, as facilities age and the tourism boom ends, tourists leave and the economy slowly declines. When the tourism boom returns due to the redevelopment of the tourist area, tourism income increases due to the large number of tourists and the economy recovers rapidly. Through these three developmental stages, an N-shaped relationship is drawn between tourism and growth. Using an economic growth model, Yang et al. (2021) derive the optimal tourism tax rate to maximize the output of the tourism industry at different stages of destination development while maintaining the carrying capacity of the destination at a sustainable level. Thus, an empirical analysis of 31 provinces in China shows that the estimated tax rates are consistent with the development stage of the tourism destination.

Lozano et al. (2008), expanding upon Gómez et al. (2008), capture the TALC by assuming a functional formation in which an incremental increase in tourists is followed by a decremental increase and finally converges to a constant number. In the initial stage of development, an increase in tourists increases revenue in the tourism sector and accelerates its growth. Moreover, owing to the small number of tourists, tourism does not significantly damage the environment. In the later stages of development, however, the congestion effect degrades the quality of the environment and reduces the amount of available public goods that contribute to production. Economic growth thus slows along with the increase of tourism revenues. Hernández and León (2013) derive the TALC as an equilibrium process rather than as a functional formation of an S-curve. In a setting where the use of physical capital degrades environmental quality, the transitional dynamic process depicts the TALC and how environmental quality changes. Other related studies include Giannoni and Maupertuis (2007), who deal with a dynamic investment model of tourism goods firms, and León et al. (2007), who analyze TALC as a social planner problem. To summarize the results of previous studies, in the later stages of development, governments could use industrial and tourism policies to improve environmental quality while transforming stagnant tourism periods into ones of rejuvenation. Innovation promoted by industrial policies can improve the carrying capacity of the destination, whereas public pollution abatement activities through tourism tax revenues can improve environmental quality (Albaladejo & Martínez-García, 2015; Cerina, 2007).

International Tourism

Globalization and Tourism

International transport accounts for 33% of global trade-related GHG emissions and more than 75% of the major industrial categories (Cristea et al., 2013). Tourism, through air transport, is currently carbon-intensive industry; therefore, the introduction of carbon taxes and emissions trading in the aviation industry, as well as the development of low-carbon technologies, are needed to achieve sustainable tourism (e.g., Lenzen et al., 2018; Scott et al., 2010; UNWTO & UNEP, 2008). It has already been highlighted by Gössling et al. (2005) and Gössling et al. (2007) that air transport for tourism has a significantly negative impact on the environment and they thus strongly advocate for the introduction of environmental regulations. Gössling and Peeters (2007) are also skeptical about carbon offsetting which was introduced by the aviation industry in response to increased regulations. The process of measuring emissions, creating an emission reduction scale, charging, and generating a certification process, which determine the efficiency and credibility of this initiative, differ from company to company, and do not provide alternative options for consumers.

Instead, McCollum et al. (2009) point out the importance of developing low-carbon technologies, such as fuel-efficient engines and lighter, more energy-efficient aircrafts, as well as of lowering costs by developing alternative fuels. Wang and Wang (2018) suggest that energy efficiency improvements can have a positive impact on tourism and environmental sustainability by reducing long-term GHG emissions in the aviation industry by around 50% compared to the case where no GHG reduction measures are implemented until 2050. Furthermore, the government should promote the development of low-carbon technologies rather than renewable energy in the tourism industry. According to Ahmad and Ma (2022), a shift from high- to low-emission tourism is 4% more effective in reducing emissions than the introduction of renewable energy in the industry.

Using an R&D-based growth model, Hamaguchi (2020) considers the following model setting: firms that also export are constantly seeking to relocate to other countries where they can earn more profit through local sales. In addition, tourists also seek to enjoy tourism not only in their own country but also by visiting destinations in other countries. Owing to the home market effect, manufacturing firms try to be in the northern portions of countries, where populations are larger, and firms can sell more products under greater demands. Exporting from south to north is costly in terms of transport, and demand in the south is relatively low. Manufacturing firms with polluting emissions are allocated grandfathered emission permits that can be traded internationally. This is a potential economic rent because the increase in the price of emission permits due to the reduction in quotas will increase the profits from the sale of emission permits at higher permit price. However, this increase in profit induces innovation and promotes economic growth via the following: firms that buy a patent for a new variety of good from a research firm can enter the industrial sector. The firms gain monopoly profits because they can sell products at a price with a mark-up due to product differentiation from others. Entrepreneurs who see the firms with these quota rents and monopoly profits try to enter this market by raising funds and buying patents. Households in the north and south are willing to finance entrepreneurs who want to buy patents by purchasing shares. Through increased R&D activities for new varieties, production expands and the economy grows as a result. This growth-promoting effect of the environmental policy is called the distribution effect. This environmental policy will lead to a concentration of firms from south to north through foreign direct investment (FDI), because entrepreneurs looking to enter a new market will seek to establish themselves via FDI in the north, where demand is greater. Although the lowering of permits will reduce global emissions, the relative amount of pollution in the north will be higher than that in the south because of the location behavior of polluting firms.

Tourism firms that produce tourism goods with labor input receive externality effects from environmental quality. This effect implies that workers' productivity increases in tourist destinations with excellent natural environments, as they can provide superior tourism services that make use of nature without the need for landscaping and preparation. Findings point to the relationship between the quality of the environment in a location and productivity output such as various services related to meals, resorts, and recreation in natural surroundings. In addition, the setting of tourist mobility is extended (Albaladejo & Martínez-García, 2015): when environmental quality is improved by a reduction in emission permits, the carrying capacity of the tourism destination also improves which leads to an increase in tourists. As a tourism policy, departure and arrival taxes are levied on the consumption of tourism goods. This tax reduction reduces tourists as well as the consumption of tourism goods, because the congestion effect reduces the carrying capacity of the destination, and additionally leads to a deterioration of the quality of the environment. The analysis shows that an improvement in environmental quality, brought about by a reduction in emission permits, will increase tourism, economic growth, and FDI, namely, tourism-led growth. Hamaguchi (2020) argues that sustainable tourism can be achieved through a mix of tourism and environmental policies, namely, an increase in departure and arrival taxes. This result is supported, to some extent, empirically: Arain et al. (2020) find a positive relationship between inbound demand and FDI, except for Mexico and Russia, and Chiu et al. (2021) identify a non-linear relationship between globalization and inbound tourism. More globalized countries attract more inbound tourists; Lee and Brahmasrene (2013), for EU member states, find that a 1% increase in FDI leads to a 0.083% increase in economic growth and a 0.017% fall in carbon dioxide emissions, and a 1% increase in FDI leads to a 0.083% increase in economic growth and a 0.017% fall in carbon emissions.

Tourism Transport and Environmental Policy

Hamaguchi (2021) considers domestic tourism, as it emits the most GHGs from short- and medium-range air transport. He argues that inter-industry emissions trading in the manufacturing and aviation sectors hinders compatibility between tourism-led growth and sustainable tourism. In this model, households derive utility from tourism by consuming aviation goods in addition to tourism goods. The aviation goods represent the distance traveled, based on the empirical result that the longer the distance traveled, and the greater they are, the greater the expenditure on tourism goods (Abbruzzo et al., 2014; Brida & Scuderi, 2013). The setting of manufacturing firms in Hamaguchi (2020) also applies to aviation firms. Development of new aeronautic technologies makes it possible to operate new routes owing to their increased productivity. By borrowing the necessary funds from households as investors who purchase technology patents, aviation firms enter new routes. Monopoly profit from exclusive routes is distributed to households as dividends. Similar to manufacturing firms, airlines also emit pollution and have, therefore, introduced emissions trading. However, they can increase their monopoly profits by selling the emission permits allocated to them by the government. Hence, the distribution effect of Hamaguchi (2020) was used. Manufacturing and aviation firms trade permits both within and between industries.

Consider a country with a strong preference for tourism over manufacturing goods and where the permits allocated to aviation is relatively low. If the emission permits of the economy remains constant, a lower emission permit of aviation leads to a relative increase in the emission permits of manufacturing firms. This leads to a fall in the demand for permits, which in turn leads to a fall in the price of permits. This reduction in production costs increases the production of aviation goods. Because of the improved quality of the environment, the production of tourism goods will also increase due to externality effects. This win-win relationship between tourism and the environment implies sustainable tourism. However, a fall in the price of emission permits leads to a decrease in the rate of economic growth through distributional effects. This fact implies that tourism-led growth will be hampered. Therefore, in countries with a strong preference for tourism goods and strict environmental regulations in the aviation industry, it is impossible to achieve both tourism-led growth and sustainable tourism. This is due to the dysfunctional nature of emissions trading, wherein a reduction in permits leads to a fall in permit prices. Alternatively, in countries with a strong preference for manufacturing goods and strict environmental regulations of manufacturing firms, the opposite is true: sustainable development, a win-win relationship between environment and growth, can be achieved, but tourism development will be hindered. Therefore, to achieve sustainable tourism, manufacturing countries such as Germany and the UK need to raise emission permits for aviation firms, whereas tourism countries such as Italy and Spain need to lower them.

Tol (2007) recommends emissions trading rather than environmental taxation. His simulation analysis shows that a carbon tax of $1000 per ton of carbon dioxide would reduce carbon dioxide emissions from international aviation by only 0.8%. By contrast, Zhang and Zhang (2018), using a simulation analysis, find that the introduction of a carbon tax would reduce carbon dioxide emissions associated with tourism, although it would result in lower GDP and reduce employment. Additionally, their policy effects differ across tourism sectors, with particularly large effects on core tourism industries such as aviation, road transport, and accommodation. Therefore, they argue that investment in carbon reduction technologies should be encouraged. The impact of environmental policies on GHG emissions from transport for tourism is not uniform and can vary depending on the policy instrument. Although environmental policy is an important instrument for achieving sustainable tourism, it should be applied with caution.

Conclusion

Sustainable Tourism Before COVID-19

Externality, congestion, and distributional effects play an important role in achieving the goal of sustainability, along with the use of environmental tax revenues to subsidize pollution abatement activities and develop low-carbon technologies. This is also true for emissions trading. However, its policy effects are being studied, and its application should be carefully considered depending on the economic structure of each country. Hamaguchi (2021), for example, and as stated above, finds that in industrialized countries such as the UK and Germany, emissions quota relaxations in the aviation sector lead to tourism development and environmental improvements, while in tourist countries such as Italy and Spain, their quota tightening leads to the opposite. The same environmental tourism policy can have different effects depending on the country and industrial structure. Further research is thus needed to determine what influences the effectiveness of such policies.

However, the discussion in this chapter of the mechanisms of sustainable tourism and its policy instruments is based on research carried out before COVID-19. This discussion can thus be seen as a potential pathway to sustainable tourism in the world before the pandemic. Changes in the social structure after COVID-19 demand a more cautious response from governments and the tourism industry. Furthermore, COVID-related deaths have, understandably, diminished the popularity and thus the profitability of tourism and other leisure industries in the years immediately following the pandemic (Lee & Chen, 2022).

In the US, tourism sub-sectors such as airlines, hotels, restaurants, and travel agencies have been affected by COVID-19 to a greater extent than other industries, with restaurants and airlines being the most affected (Abdelsalam et al., 2021). Wang et al. (2022) use China as a model case and state that relaxed policies could result in a 97% loss of inbound demand revenue, an 11% fall in its real GDP, and a 15% fall in its employment. Payne et al. (2022) see this impact as permanent in the Croatian tourism industry. However, there is a V-shaped relationship between COVID recoveries and profitability of the tourism and leisure industry (Lee & Chen, 2022), and appropriate intervention control policies in China will mitigate the impact of COVID-19 on tourism and non-tourism industries (Wang et al., 2022). Using key macroeconomic variables, Plzáková and Smeral (2022) outline a detailed forecast scenario for international tourism demand within the European Union in the tourism import balance.

Policies focusing on reviving the tourism industry will contribute to the rapid recovery of the economy but will in turn have serious environmental consequences. Thus, as Chakraborty and Maity (2020) note, the pandemic led to an across-the-board decline in GDP and a sharp rise in unemployment in many countries, while the lockdown policy contributed to a reduction in GHGs.

The stagnation of economic activity, such as reduced traffic and halted production, resulted in reduced emissions of pollutants and factory effluents. Even in countries where the disease's spread, such in as China, the USA, Italy, and Spain, led to the declaration of a state of emergency, COVID-19 has resulted in improved air quality, the beautification of beaches, and to reduced noise for some time, which in turn leads to higher levels of population health (Zambrano-Monserrate et al., 2020). Li et al. (2022) suggest that quarantine policies are effective in both preventing the spread of COVID-19 and developing tourism. Zambrano-Monserrate et al. (2020) are concerned that a gradual return of global economic activity to pre-COVID levels could lead to a rise in short-term GHG emissions above pre-COVID levels. This is concerning because "revenge consumption" by consumers holding back on travel may lead to a V-shaped recovery in the tourism industry. Measures to support the tourism industry will add to this trend.

Policy Mix After COVID-19

As Gössling et al. (2021) argue, the current model of tourism is unsustainable environmentally and makes possible the rapid and global spread of illness. A boom in global tourism pre-pandemic facilitated the spread of disease. Tourists enter into contact with large numbers of people and participate in activities in restaurants, theaters, and other venues where, because food and drink are being consumed and people are projecting their voices or conversing, diseases can easily be transmitted through droplets. Thus, as Gössling et al. (2021) point out, new measures are needed to make tourism more environmentally sustainable and to prevent the rapid spread of diseases such as COVID-19. Furthermore, the rapid recovery of the tourism industry as countries around the world reopened will have significant consequences to the environment. Before the recent/current pandemic, the world had been working toward sustainable tourism by introducing emissions trading and carbon taxes, in addition to promoting green innovation in the aviation industry. However, simply imposing stringent environmental regulations may not be sufficient to cope with the massive GHG emissions caused by the rapid recovery of the tourism industry. In addition, it is necessary to deal with tourism workers who have suffered economic damage due to COVID-19.

The tourism industry comprises a few large enterprises with land and capital, as well as small- and medium-sized enterprises and family businesses with limited financial resources. It is a labor-intensive industry; for instance, hotels and other accommodation facilities employ many workers for various tasks. Moreover, as there are distinct busy and off-peak periods, many workers are employed temporarily during peak seasons. These fluctuations lead companies to secure profits through employment adjustments rather than to invest in long-term workers. Vu and Hartley (2022) find that the growth in tourism from 2000 to 2015 is mainly due to an increase in hired labor, while labor productivity has been declining in most countries. Since labor productivity determines workers' wages, the economic environment for workers in the tourism industry has been deteriorating since before the pandemic. The contraction of the tourism industry and increase in unemployment caused by COVID-19 have accelerated this trend. These situations are exacerbated by natural disasters caused by climate change. Frequent hurricanes destroy touristic infrastructure, which in turn causes an increase in unemployment. Even after recovery, large employers will curtail the hiring of workers until a certain number of tourists return. Non-regular workers who were dependent on employment during peak periods will be impoverished due to prolonged unemployment. Micro-enterprises with limited financial resources may choose to close their businesses, unable to provide the necessary funds for recovery. Furthermore, many tourism industry workers live near touristic destinations and will therefore be personally affected by the same hurricanes which have impacted their livelihood.

Moreover, workers in the tourism industry may find it difficult to secure jobs in other industries due to an array of barriers. This further widens the enterprise gap between large and micro-enterprises and the economic gap between the employed and unemployed.

A policy mix with income redistribution policies is important for tourism sustainability, as measures to promote tourism after COVID-19 may increase inequality. Vu and Hartley (2022) suggest that the tourism sector could recover if it improves labor productivity in anticipation of the end of the pandemic. Sustainable tourism after COVID-19 can be achieved through a policy mix.

Without the necessary adaptation measures, air travel in tourism will continue to emit significant amounts of greenhouse gases. However, the tourism industry is both dependent on the natural environment and vulnerable to natural disasters. Moreover, as the industry is labor-intensive, the destruction of tourism due to natural disasters is directly linked to income inequality. The various effects of tourism on the economy and the environment, as identified by previous studies, should be taken into account when discussing policy measures to contribute to their resolution.

References

Abbruzzo, A., Brida, J. G., & Scuderi, R. (2014). Determinants of individual tourist expenditure as a network: Empirical findings from Uruguay. *Tourism Management, 43*, 36–45. https://doi.org/10.1016/j.tourman.2014.01.014

Abdelsalam, O., Aysan, A. F., Cepni, O., & Disli, M. (2021). The spillover effects of the COVID-19 pandemic: Which subsectors of tourism have been affected more? *Tourism Economics, 29*(2), 559–567. https://doi.org/10.1177/13548166211053670

Adedoyin, F. F., Seetaram, N., Disegna, M., & Filis, G. (2021). The effect of tourism taxation on international arrivals to a small tourism-dependent economy. *Journal of Travel Research, 62*(1), 135–153. https://doi.org/10.1177/00472875211053658

Ahmad, N., & Ma, X. (2022). How does tourism development affect environmental pollution? *Tourism Economics, 28*(6), 1453–1479. https://doi.org/10.1177/13548166211000480

Albaladejo, I. P., & Martínez-García, M. P. (2015). An R&D-based endogenous growth model of international tourism. *Tourism Economics, 21*(4), 701–719. https://doi.org/10.5367/te.2014.0379

Antonakakis, N., Dragouni, M., Eeckels, B., & Filis, G. (2019). The tourism and economic growth enigma: Examining an ambiguous relationship through multiple prisms. *Journal of Travel Research, 58*(1), 3–24. https://doi.org/10.1177/0047287517744671

Arain, H., Han, L., Sharif, A., & Meo, M. S. (2020). Investigating the effect of inbound tourism on FDI: The importance of quantile estimations. *Tourism Economics, 26*(4), 682–703. https://doi.org/10.1177/1354816619859695

Balagner, J., & Cantavela-Jordá, M. (2002). Tourism as a long-run economic factor: The Spanish case. *Applied Economics, 34*(7), 877–884. https://doi.org/10.1080/00036840110058923

Barrera, F., & Garrido, N. (2018). Public holidays, tourism, and economic growth. *Tourism Economics, 24*(4), 473–485. https://doi.org/10.1177/1354816617749340

Brida, J. G., Cortes-Jimenez, I., & Pulina, M. (2016). Has the tourism-led growth hypothesis been validated? A literature review. *Current Issues in Tourism, 19*(5), 394–430. https://doi.org/10.1080/13683500.2013.868414

Brida, J. G., & Scuderi, R. (2013). Determinants of tourist expenditure: A review of microeconometric models. *Tourism Management Perspectives, 6*, 28–40. https://doi.org/10.1016/j.tmp.2012.10.006

Buckley, R. (2011). Tourism and environment. *Annual Review of Environment and Resources, 36*(1), 397–416. https://doi.org/10.1146/annurev-environ-041210-132637

Butler, R. W. (1980). The concept of a tourist area cycle of evolution: Implications for management of resources. *The Canadian Geographer/Le Géographe Canadien, 24*(1), 5–12. https://doi.org/10.1111/j.1541-0064.1980.tb00970.x

Cárdenas-García, P. J., Sánchez-Rivero, M., & Pulido-Fernández, J. I. (2015). Does tourism growth influence economic development? *Journal of Travel Research, 54*(2), 206–221. https://doi.org/10.1177/0047287513514297

Cerina, F. (2007). Tourism specialization and environmental sustainability in a dynamic economy. *Tourism Economics, 13*(4), 553–582. https://doi.org/10.5367/000000007782696032

Cerina, F., Markandya, A., & McAleer, M. (2010). *Economics of sustainable tourism*. Routledge.

Chakraborty, I., & Maity, P. (2020). COVID-19 outbreak: Migration, effects on society, global environment and prevention. *Science of The Total Environment, 728*, 138882. https://doi.org/10.1016/j.scitotenv.2020.138882

Chiu, Y.-B., & Yeh, L.-T. (2017). The threshold effects of the tourism-led growth hypothesis: Evidence from a cross-sectional model. *Journal of Travel Research, 56*(5), 625–637. https://doi.org/10.1177/0047287516650938

Chiu, Y.-B., Zhang, W., & Ding, K. (2021). Does globalization influence inbound tourism? Evidence from a dynamic panel threshold analysis. *Journal of Travel Research, 60*(5), 1074–1084. https://doi.org/10.1177/0047287520919514

Cristea, A., Hummels, D., Puzzello, L., & Avetisyan, M. (2013). Trade and the greenhouse gas emissions from international freight transport. *Journal of Environmental Economics and Management, 65*(1), 153–173. https://doi.org/10.1016/j.jeem.2012.06.002

Croes, R., Ridderstaat, J., & van Niekerk, M. (2018). Connecting quality of life, tourism specialization, and economic growth in small island destinations: The case of Malta. *Tourism Management, 65*, 212–223. https://doi.org/10.1016/j.tourman.2017.10.010

Das, J., & Dirienzo, C. (2010). Tourism competitiveness and corruption: A cross-country analysis. *Tourism Economics, 16*(3), 477–492. https://doi.org/10.5367/000000010792278392

Deng, T., Ma, M., & Cao, J. (2014). Tourism resource development and long-term economic growth: A resource curse hypothesis approach. *Tourism Economics, 20*(5), 923–938. https://doi.org/10.5367/te.2013.0325

European Commission. (2017). *The EU tackles growing aviation emissions*. http://europa.eu/rapid/press-releaseIP-17-189en.htm

Giannoni, S., & Maupertuis, M.-A. (2007). Environmental quality and optimal investment in tourism infrastructures: A small Island perspective. *Tourism Economics, 13*(4), 499–513. https://doi.org/10.5367/000000007782696014

Gómez, C. M., Lozano, J., & Rey-Maquieira, J. (2008). Environmental policy and long- term welfare in a tourism economy. *Spanish Economic Review, 10*(1), 41–62. https://doi.org/10.1007/s10108-007-9028-0

Gössling, S., Broderick, J., Upham, P., Ceron, J.-P., Dubois, G., Peeters, P., & Strasdas, W. (2007). Voluntary carbon offsetting schemes for aviation: Efficiency, credibility and sustainable tourism. *Journal of Sustainable Tourism, 15*(3), 223–248. https://doi.org/10.2167/jost758.0

Gössling, S., & Peeters, P. (2007). 'It does not harm the environment!' An analysis of industry discourses on tourism, air travel and the environment. *Journal of Sustainable Tourism, 15*(4), 402–417. https://doi.org/10.2167/jost672.0

Gössling, S., Peeters, P., Ceron, J.-P., Dubois, G., Patterson, T., & Richardson, R. B. (2005). The eco-efficiency of tourism. *Ecological Economics, 54*(4), 417–434. https://doi.org/10.1016/j.ecolecon.2004.10.006

Gössling, S., Scott, D., & Hall, C. M. (2021). Pandemics, tourism and global change: A rapid assessment of COVID-19. *Journal of Sustainable Tourism, 29*(1), 1–20. https://doi.org/10.1080/09669582.2020.1758708

Hamaguchi, Y. (2020). Do pollution havens restrict tourism-led growth? Achieving sustainable tourism via a mix of environmental and tourism policies. *Tourism Economics, 26*(7), 1175–1196. https://doi.org/10.1177/1354816619868086

Hamaguchi, Y. (2021). Does the trade of aviation emission permits lead to tourism-led growth and sustainable tourism? *Transport Policy, 105*, 181–192. https://doi.org/10.1016/j.tranpol.2021.03.012

Hernández, J. M., & León, C. J. (2013). Welfare and environmental degradation in a tourism-based economy. *Tourism Economics, 19*(1), 5–35. https://doi.org/10.5367/te.2013.0191

Hjalager, A.-M. (2010). A review of innovation research in tourism. *Tourism Management, 31*(1), 1–12. https://doi.org/10.1016/j.tourman.2009.08.012

Johnston, R. J., & Tyrrell, T. J. (2005). A dynamic model of sustainable tourism. *Journal of Travel Research, 44*(2), 124–134. https://doi.org/10.1177/0047287505278987

Lee, C.-C., & Chang, C.-P. (2008). Tourism development and economic growth: A closer look at panels. *Tourism Management, 29*(1), 180–192. https://doi.org/10.1016/j.tourman.2007.02.013

Lee, C.-C., & Chen, M.-P. (2022). The impact of COVID-19 on the travel and leisure industry returns: Some international evidence. *Tourism Economics, 28*(2), 451–472. https://doi.org/10.1177/1354816620971981

Lee, J. W., & Brahmasrene, T. (2013). Investigating the influence of tourism on economic growth and carbon emissions: Evidence from panel analysis of the European Union. *Tourism Management, 38*, 69–76. https://doi.org/10.1016/j.tourman.2013.02.016

Lee, R., Lockshin, L., Cohen, J., & Corsi, A. (2019). A latent growth model of destination image's halo effect. *Annals of Tourism Research, 79*, 102767. https://doi.org/10.1016/j.annals.2019.102767
Lenzen, M., Sun, Y.-Y., Faturay, F., Ting, Y.-P., Geschke, A., & Malik, A. (2018). The carbon footprint of global tourism. *Nature Climate Change, 8*(6), 522–528. https://doi.org/10.1038/s41558-018-0141-x
León, C. J., Hernández, J. M., & González, M. (2007). Economic welfare, the environment and the tourist product life cycle. *Tourism Economics, 13*(4), 583–601. https://doi.org/10.5367/000000007782696078
Li, S., Ma, S., & Zhang, J. (2022). Building a system dynamics model to analyze scenarios of COVID-19 policymaking in tourism-dependent developing countries: A case study of Cambodia. *Tourism Economics, 29*(2), 488–512. https://doi.org/10.1177/13548166211059080
Liu, A., & Wu, D. C. (2019). Tourism productivity and economic growth. *Annals of Tourism Research, 76*, 253–265. https://doi.org/10.1016/j.annals.2019.04.005
Liu, Z. (2003). Sustainable tourism development: A critique. *Journal of Sustainable Tourism, 11*(6), 459–475. https://doi.org/10.1080/09669580308667216
Lozano Ibáñez, J., Rey-Maquieira, J., & Gomez, C. (2004). Land, environmental externalities and tourism development. *SSRN Electronic Journal*. https://doi.org/10.2139/ssrn.504423
Lozano, J., Gómez, C. M., & Rey-Maquieira, J. (2008). The TALC hypothesis and economic growth theory. *Tourism Economics, 14*(4), 727–749. https://doi.org/10.5367/000000008786440166
Marsiglio, S. (2015). Economic growth and environment: Tourism as a trigger for green growth. *Tourism Economics, 21*(1), 183–204. https://doi.org/10.5367/te.2014.0411
McCollum, D., Gould, G., & Greene, D. (2009). *Greenhouse gas emissions from aviation and marine transportation: Mitigation potential and policies*. Pew Center on Global Climate Change.
Nunkoo, R., Seetanah, B., Jaffur, Z. R. K., Moraghen, P. G. W., & Sannassee, R. V. (2020). Tourism and economic growth: A meta-regression analysis. *Journal of Travel Research, 59*(3), 404–423. https://doi.org/10.1177/0047287519844833
Ouattara, B., Pérez-Barahona, A., & Strobl, E. (2019). Dynamic implications of tourism and environmental quality: OUATTARA et al. *Journal of Public Economic Theory, 21*(2), 241–264. https://doi.org/10.1111/jpet.12330

Papavasileiou, E. F., & Tzouvanas, P. (2021). Tourism Carbon Kuznets-Curve hypothesis: A systematic literature review and a paradigm shift to a corporation-performance perspective. *Journal of Travel Research, 60*(4), 896–911. https://doi.org/10.1177/0047287520915276

Paramati, S. R., Alam, M. S., & Chen, C.-F. (2017). The effects of tourism on economic growth and CO_2 emissions: A comparison between developed and developing economies. *Journal of Travel Research, 56*(6), 712–724. https://doi.org/10.1177/0047287516667848

Parrilla, J. C., Font, A. R., & Nadal, J. R. (2007). Tourism and long-term growth a Spanish perspective. *Annals of Tourism Research, 34*(3), 709–726. https://doi.org/10.1016/j.annals.2007.02.003

Payne, J. E., Gil-Alana, L. A., & Mervar, A. (2022). Persistence in Croatian tourism: The impact of COVID-19. *Tourism Economics, 28*(6), 1676–1682. https://doi.org/10.1177/1354816621999969

Pina, I. P. A. L., & Martínez-García, M. P. (2013). An endogenous growth model of international tourism. *Tourism Economics, 19*(3), 509–529. https://doi.org/10.5367/te.2013.0212

Plzáková, L., & Smeral, E. (2022). Impact of the COVID-19 crisis on European tourism. *Tourism Economics, 28*(1), 91–109. https://doi.org/10.1177/13548166211031113

Pulido-Fernández, J. I., & Cárdenas-García, P. J. (2021). Analyzing the bidirectional relationship between tourism growth and economic development. *Journal of Travel Research, 60*(3), 583–602. https://doi.org/10.1177/0047287520922316

Romão, J. (2020). Tourism, smart specialisation, growth, and resilience. *Annals of Tourism Research, 84*, 102995. https://doi.org/10.1016/j.annals.2020.102995

Schubert, S. F., & Schamel, G. (2021). Sustainable tourism development: A dynamic model incorporating resident spillovers. *Tourism Economics, 27*(7), 1561–1587. https://doi.org/10.1177/1354816620934552

Scott, D., Peeters, P., & Gössling, S. (2010). Can tourism deliver its "aspirational" greenhouse gas emission reduction targets? *Journal of Sustainable Tourism, 18*(3), 393–408. https://doi.org/10.1080/09669581003653542

Song, H., & Wu, D. C. (2022). A critique of tourism-led economic growth studies. *Journal of Travel Research, 61*(4), 719–729. https://doi.org/10.1177/00472875211018514

Tol, R. S. J. (2007). The impact of a carbon tax on international tourism. *Transportation Research Part D: Transport and Environment, 12*(2), 129–142. https://doi.org/10.1016/j.trd.2007.01.004

Tugcu, C. T. (2014). Tourism and economic growth nexus revisited: A panel causality analysis for the case of the Mediterranean Region. *Tourism Management, 42*, 207–212. https://doi.org/10.1016/j.tourman.2013.12.007
UNWTO & UNEP. (2008). *Climate change and tourism-responding to global challenges*. UNWTO and UNEP.
Vu, K., & Hartley, K. (2022). Drivers of growth and catch-up in the tourism sector of industrialized economies. *Journal of Travel Research, 61*(5), 1156–1172. https://doi.org/10.1177/00472875211019478
Wang, C., Meng, X., Siriwardana, M., & Pham, T. (2022). The impact of COVID-19 on the Chinese tourism industry. *Tourism Economics, 28*(1), 131–152. https://doi.org/10.1177/1354816621104l209
Wang, M.-C., & Wang, C.-S. (2018). Tourism, the environment, and energy policies. *Tourism Economics, 24*(7), 821–838. https://doi.org/10.1177/1354816618781458
World Tourism Organization (UNWTO). (2015). Tourism and the sustainable development goals. https://doi.org/10.18111/9789284417254
Yang, C., Lin, Z., Huang, J., & Cheng, T. (2021). A dynamic tax model based on destination lifecycle for sustainable tourism development. *Journal of Travel Research*, 00472875211057596. https://doi.org/10.1177/00472875211057596
Zaman, K., Shahbaz, M., Loganathan, N., & Raza, S. A. (2016). Tourism development, energy consumption and Environmental Kuznets Curve: Trivariate analysis in the panel of developed and developing countries. *Tourism Management, 54*, 275–283. https://doi.org/10.1016/j.tourman.2015.12.001
Zambrano-Monserrate, M. A., Ruano, M. A., & Sanchez-Alcalde, L. (2020). Indirect effects of COVID-19 on the environment. *Science of the Total Environment, 728*, 138813. https://doi.org/10.1016/j.scitotenv.2020.138813
Zhang, J., & Zhang, Y. (2018). Carbon tax, tourism CO2 emissions and economic welfare. *Annals of Tourism Research, 69*, 18–30. https://doi.org/10.1016/j.annals.2017.12.009
Zhang, J., & Zhang, Y. (2021). Tourism, economic growth, energy consumption, and CO2 emissions in China. *Tourism Economics, 27*(5), 1060–1080. https://doi.org/10.1177/1354816620918458
Zuo, B., & Huang, S. S. (2018). Revisiting the tourism-led economic growth hypothesis: The case of China. *Journal of Travel Research, 57*(2), 151–163. https://doi.org/10.1177/0047287516686725

12

Can Tourist Taxes Make Tourism More Sustainable?

José Luis Durán-Román,
Juan Ignacio Pulido-Fernández,
and Yaiza López-Sánchez

Introduction

The tourism industry was hit particularly hard by the COVID-19 pandemic. The number of international tourists dropped by 74% in 2020 and jeopardized between 100 and 120 million direct jobs worldwide (United Nations World Tourism Organization [UNWTO], 2021), a dramatic turn of events that sent tourist activity back to levels not seen since the 1990s. However, with every crisis or catastrophe comes opportunities for improvement or the chance to begin anew.

The pandemic sparked an increase in demand for destinations that are less crowded, more sustainable, and more environmentally conscious. While, according to Corbisiero and Monaco (2021), tourist behavior is influenced by a number of factors, such as personal economic well-being,

J. L. Durán-Román
Department of Business Organization, Marketing and Sociology,
University of Jaén, Jaén, Spain
e-mail: jduran@ujaen.es

T. Walker et al. (eds.), *Sustainable Tourism*,
https://doi.org/10.1007/978-3-031-43528-7_12

costs, perceived health risks, and so on, it is equally true that interest in sustainable tourism is also increasingly important (Nelson et al., 2021). The term "sustainable tourism" began to be used by academic researchers at the beginning of the 1990s, as an application of the general concept of sustainable development, popularized by the Brundtland report. According to the UNWTO (n.d.), sustainable tourism is a "tourism that takes full account of its current and future economic, social and environmental impacts, addressing the needs of visitors, the industry, the environment, and host communities."

In parallel, from the perspective of supply, and coinciding with the boom in tourism experienced since the mid-twentieth century, many countries—mainly around the Mediterranean—have opted to specialize in "beach" tourism, concentrating supply on coastlines, and drawing large crowds of tourists at very specific times of the year.

Spain is a clear example of a country that has specialized in this type of tourism (Gutiérrez et al., 2018). Of the 31.18 million international tourists who visited the country in 2021, 82.22% came to five regions characterized by their specialization in coastal tourism (Andalusia, the Canary Islands, Catalonia, the Balearic Islands, and the Region of Valencia) (Instituto Nacional de Estadística [INE], 2022b). Although there are certain limitations when it comes to conceptualizing these tourist regions as uniform entities, based exclusively on the model of beach tourism (since

J. I. Pulido-Fernández (✉)
Laboratory of Analysis and Innovation in Tourism (LAInnTUR), University of Jaén, Jaén, Spain

Department of Applied Economics, University of Jaén, Jaén, Spain
e-mail: jipulido@ujaen.es

Y. López-Sánchez
Laboratory of Analysis and Innovation in Tourism (LAInnTUR), University of Jaén, Jaén, Spain

Department of Applied Economics (Political economy), University of Málaga, Málaga, Spain
e-mail: ylopez@uma.es

they have other resources and attractions), the highest percentage of spending among tourists visiting Spain has been linked to beach tourism (INE, 2022a).

The model of beach tourism followed by Spain and other countries has been showing major signs of exhaustion since the start of this century (Aguiló et al., 2005; Benner, 2020). In addition, this model has resulted in a significant negative impact on the territory, since many of the economic benefits have been obtained at the expense of environmental and socio-cultural balance (Drius et al., 2019). This has led to the degradation of fragile coastal resources (López-Sánchez & Pulido-Fernández, 2014), which translates into a decline in the attractiveness of the destination and its competitiveness, as well as a loss of well-being for the resident population.

The maturity and decline of major tourist destinations brings challenges that arise from changes in travel habits and patterns of tourist behavior (Benner, 2020). These changes require different models of tourist specialization capable of transforming the relationship between tourism and nature, climate and the economy, and moving toward a more sustainable, inclusive, and resilient form of tourism.

Alternatively, it is also possible to rejuvenate a tourist destination, although this only results in a short-term solution unless a profound change is made in the attractions on which the destination is based (Butler, 1980). Indeed, mature destinations, especially coastal ones, must continuously renew and offer new solutions to overcome the sustainability problems they are facing. "Tourist destinations can be flexible in time, with the ability to continuously adapt to global changes with specific local responses," reconfiguring the product and targeting new markets and segments (Chapman & Light, 2016; Gomis-López & González-Reverté, 2020, p. 4).

Many destinations currently in the stages of maturity and/or decline, based on an unsustainable model of mass tourism, depend more than ever on tourism policies committed to sustainable management (Gutiérrez et al., 2018). In fact, a commitment to sustainability becomes an important element of differentiation and competitiveness in tourist markets (Crespi-Vallbona & Mascarilla-Miró, 2021). The transition to sustainable tourism, however, is a difficult task and requires time, money, and a willingness among stakeholders to undertake the necessary changes and implement sustainable practices (Simeoni et al., 2019).

The development of new public policies aiming to reposition a destination on the basis of sustainability requires an allocation of funds, which, in most cases, exceeds public budgets (Mansourian & Dudley, 2008). It is, therefore, necessary to develop new strategies that allow for the generation of new sources of public income. Along these lines, the issue of financing tourist destinations and the search for new revenues has been an emerging global issue since at least 1980 (Mak, 2006). At that time there was already a major debate about the unchecked rise in public spending on tourism, requiring governments to explore new sources of supplementary funding.

Clearly, the funding system at any level of government should be designed to take into account the powers transferred to it with regard to tourism. This requires new funding formulas that, from a position of equity, fall to the agents involved in the tourism market and which make it possible to cover the public spending commitments of public authorities (Pulido & Cárdenas, 2012). Furthermore, policies related to the sustainability and competitiveness of the tourist destination are also required. Indeed, inadequate funding is one of the greatest obstacles to the development of the destination market (Wilson et al., 2001).

Having, therefore, justified the need to transition toward a more competitive and sustainable model of tourism, especially in mature beach tourist destinations, the key issue, then, is to determine the sources of income with which to finance the actions needed to achieve this goal. On this basis, one of the main challenges facing public administrations, mainly in areas with a strong focus on tourism, is to identify new funding formulas.

Fiscal measures—taxes and fees—levied on tourist activity are instruments of economic and fiscal policy that contribute to the collection of generalized public revenue (Banús, 2021) and, in turn, facilitate the development of public policies to design a more sustainable and differentiated supply of tourism at the destination. In addition, per Adedoyin et al. (2021): "taxation is often the instrument that is used to internalize the external cost to reduce demand to a more socially acceptable level" (p. 4).

In this search for new funding formulas, a large body of research has examined tourist willingness to pay (WTP) in order to improve the competitiveness and sustainability of the tourist destination. For example, Birdir et al. (2013) analyze around 20 research papers that explore tourist WTP in coastal destinations. However, little research has analyzed the WTP of tourism businesses (Bernard et al., 2009; Lan et al., 2014; Mäntymaa et al., 2019). In addition, to the best of our knowledge, no previous studies have used a mixed approach to WTP, examining the segmentation of tourism businesses, together with the establishment of tourist taxes, which is the novel contribution of this chapter.

The aim of this chapter is to improve understanding of how tourist businesses behave in the context of mature tourist destinations, by grouping them into homogeneous segments or subsets, identifying what factors explain willingness to pay (WTP) in each segment to achieve a more sustainable and competitive destination. The effectiveness of policy formulation and implementation aimed at restructuring tourism models, based on sustainable management and innovation, and as key market elements improve the competitiveness of the destination, will largely depend on understanding the perception of tourist businesses regarding their WTP tourist taxes.

In addition, before establishing any tax instruments, to ensure higher levels of acceptance (Lindsey & Holmes, 2010; Mazzocchi et al., 2019), policymakers should know the WTP of involved agents—tourists and/or businesses—and their preferred form of payment.

The study described here was carried out in Andalusia, a major tourist destination in southern Spain that attracted 20.05 million tourists in 2021 (Instituto de Estadística y Cartografía de Andalucía [IECA], 2022). As a destination specializing in mass tourism, it has reached maturity and has, for decades, been showing signs of stagnation. The study consisted of 916 surveys completed by tourism businesses in Andalusia, to which we applied a two-step cluster analysis.

We formulate the following hypotheses: (1) there are one or more segments of the tourism business that demonstrate different WTP taxes and/or fees with regard to their commitment to the sustainability of the destination and improved competitiveness, and (2) there are certain tax instruments that are more widely accepted than others among tourism businesses in Andalusia.

The remainder of the study is structured as follows: the "Methodology" section describes the methodology used and the rationale behind the choice of statistical technique selected. The "Results" section presents the results of the study, while "Conclusions" presents the main conclusions as well as a discussion of the limitations of the study.

Methodology

Questionnaire and Data Sheet

To verify the hypotheses posited by this study, we conducted a survey of tourism businesses in Andalusia (Table 12.1). The survey was structured as follows: the first block classified tourism businesses in Andalusia according to the main business variables (type of business, size of business, and annual turnover). The second block, using Likert-type scales (scored from 1 to 7), contained questions about the competitiveness of tourism, sustainability, and the possibility of improving the tourist experience in Andalusia. The third block identified WTP among Andalusia's tourism businesses, in order to make the destination more sustainable, more competitive, and to improve the experiences of tourists visiting Andalusia. Open, closed (dichotomous and multiple choice), and mixed questions were used. In addition, nominal, ordinal, and Likert-type scales were used to measure the variables under examination.

Table 12.1 Technical details of the questionnaire

Universe	Tourist businesses in Andalusia
Scope of study	Andalusia
Type of questionnaire	Structured questionnaire conducted by personal telephone interview
Sample Size	916 valid surveys
Sampling error	± 3.1%
Level of confidence	95% ($p = q = 0.50$)
Period of fieldwork	July to September 2019

The questions included in this third block analyze the willingness of tourism businesses to pay six specific types of taxation levied on different aspects related to tourist activity: (1) tax on overnight stays in P2P accommodation; (2) waste disposal tax for tourism businesses; (3) charge for the use of facilities; (4) tax on passenger boats and ships visiting protected natural areas; (5) tax on activities in protected areas; and (6) environmental conservation tax in municipalities whose main activity is eco-tourism. These types of taxation were defined earlier by a panel of experts assembled for a previous study. Having selected the sample size, stratified quota sampling was completed, respecting the population weightings of each stratum, and using simple random sampling of each stratum.

Table 12.2 shows descriptive information about the sample analyzed: the total distribution of interviews conducted with Andalusian tourism businesses, in accordance with the classification defined by Andalusia's Regional Department of Tourism, Regeneration, Justice and Local Administration, as well as the proportion of tourism businesses classified in each category, in accordance with the official database provided by the Department for the purposes of this study.

Table 12.2 Characterization of Andalusian tourist businesses (sample analyzed)

	Number	% of table *N*
Activity	**916**	**100**
Travel agencies	229	25
Accommodation	1	0.1
Rural accommodation	343	37.4
Holiday apartments	189	20.6
Camping	24	2.6
Active tourism, leisure, and adventure activities	130	14.2
Size (number of employees)	**916**	**100**
Self-employed/no employees	260	28.4
1–9	584	63.8
10–49	65	7.1
50–249	7	0.8
Turnover tranche (€)	**916**	**100**
Up to 50,000	221	24.1
Between 50,001 and 100,000	231	25.2
Between 100,001 and 500,000	406	44.3
Between 500,001 and 1,000,000	45	4.9
Between 1,000,001 and 2,000,000	8	0.9
More than 2,000,000	5	0.5

Note: Due to rounding, the column totals may not always add up to 100%

Analytical Model

According to scientific literature in the field of tourism, segmentation involves dividing the market into groups with similar behaviors (Levin & Zahavi, 2001). For the purposes of this study, we employ a two-step cluster analysis (Chiu et al., 2001).

Two-step cluster analysis is an exploratory tool designed to discover the natural groupings—or clusters—of a dataset that would otherwise not be detectable (Pérez, 2011). The algorithm is applied in two stages. In the first stage, the initial input data are compressed into a manageable set of sub-clusters. The second stage uses a hierarchical clustering method to progressively merge sub-clusters into ever-larger clusters, without having to run the data again.

This algorithm has certain advantages over traditional clustering techniques (Pérez, 2011):

- Treatment of categorical and continuous variables: This procedure works with both continuous and categorical variables. Cases represent the objects to group into clusters, and the variables represent the attributes on which clustering is based. Assuming that the variables are independent, it is possible to apply a joint multinomial normal distribution to continuous and categorical variables.
- Automatic selection of the number of clusters: By comparing the values of a model selection criterion for different clustering solutions, the procedure can automatically determine the optimal number of clusters.
- Scalability: By constructing a cluster feature (CF) tree that summarizes the entries, the two-step algorithm can analyze large data files.
- Exclusion of anomalies: It can be configured to automatically exclude outliers or very strange cases that may contaminate the results.

Results

In order to be able to analyze the relationship with WTP *a posteriori*, the study explores the underlying typologies or groupings of tourist businesses. Following a prior exploration process in which different combinations of segmental factors were tested, an optimal model was achieved—100% of the sample—and the following variables are found to possess the greatest discriminatory potential: business size, activity, and turnover. In addition, three clusters with sufficient data quality were obtained.

- Cluster 1 (37.9% of the sample) labeled "micro-enterprises": this cluster groups businesses that employ between one and nine workers, operating mainly as travel agencies and rural accommodation providers, with an annual average turnover of €190,066.
- Cluster 2 (33.7% of the sample) labeled "SMEs" (small- and medium-sized enterprises): this cluster groups businesses with more than nine workers, as well as large companies (more than 250 employees). This cluster is mainly operating in the following fields: apartments, camping, and active tourism businesses, with an average turnover of €324,631.
- Cluster 3 (28.4% of the sample) labeled "sole traders": this cluster groups tourist businesses that have no employees, whose activity focuses on rural accommodation and travel agencies, with an average turnover of €55,017.

Tourism, Competitiveness, and Investment According to Business Profile

To explore the characteristics of each of the three clusters identified in greater depth, and prior to identifying the WTP of these segments, a number of questions were asked regarding the costs arising from the provision of certain public goods and services, the impacts generated by tourism, and the need for new investments to make the destination more competitive (Table 12.3).

The results in Table 12.3 show a high degree of agreement among the informants about the three issues. More specifically, the three clusters agree that public spending and costs are higher due to the provision of certain kinds of infrastructure and public services for tourism (5.73), that tourist activity generates a number of negative impacts (5.56), and that making the destination more competitive involves improving

Table 12.3 Tourism, competitiveness, and investment

		N Valid	Mean	Standard deviation
Do you think that both the costs of maintaining infrastructure and the costs of providing public services are higher due to tourism?	Total	916	5.73	1.08
	Sole traders	260	5.30	1.31
	Micro-enterprises	347	5.90	0.97
	SMEs	309	5.91	0.87
Do you think that tourism has negative impacts on the tourist destination? For example, waste generation, pollution, congestion at tourist sites, vandalism, saturation of certain services, etc.?	Total	916	5.56	1.34
	Sole traders	260	5.05	1.53
	Micro-enterprises	347	5.71	1.30
	SMEs	309	5.83	1.08
Do you think that, in order to make the destination more competitive, existing infrastructures and services should be improved?	Total	916	6.13	0.86
	Sole traders	260	5.91	0.91
	Micro-enterprises	347	6.21	0.92
	SMEs	309	6.23	0.71

infrastructure and services (6.13). Although the highest average value of the three issues is obtained in the SME segment (5.91, 5.83, and 6.23, respectively), the p-value of the Kruskal-Wallis (K-W) test is 0.000 for the three issues increased. There are, therefore, significant differences in the opinions of the three segments. The post-hoc pair comparison tests indicate that the scores for the three issues in segment 1 (sole traders) are significantly lower than the other two clusters.

The participants were then asked which options could improve the competitiveness of tourism in Andalusia. As Table 12.4 shows, the improvement of tourism infrastructure (5.95) is the main option through which tourism could be made more competitive, followed by tourist services (5.85), and finally general infrastructure (5.84). The Kruskal-Wallis tests, with their p-values of 0.006, 0.087, and 0.002, respectively, indicate that, except for the tourist infrastructure option, there are significant differences between segments for the remainder. Sole traders obtained a lower score than the other segments for general infrastructure and tourist services.

Table 12.4 Options for improving the competitiveness of tourism in Andalusia

		N Valid	Mean	Standard deviation
General infrastructure (public transport, safety, cleaning, congestion, traffic, etc.)	Total	916	5.84	1.14
	Sole traders	260	5.67	1.25
	Micro-enterprises	347	5.83	1.22
	SMEs	309	6.01	0.92
Tourist infrastructure (conservation and maintenance of tourist attractions, emblematic buildings, environment, etc.)	Total	916	5.95	1.14
	Sole traders	260	5.86	1.15
	Micro-enterprises	347	5.99	1.20
	SMEs	309	5.99	1.06
Tourist services (cultural and leisure offer, tourist accommodation, tourist information services, internet connection, etc.)	Total	916	5.85	1.18
	Sole traders	260	5.65	1.28
	Micro-enterprises	347	5.94	1.24
	SMEs	309	5.92	0.99

Table 12.5 Willingness to pay among Andalusian tourist businesses

		Sole traders		Micro-enterprises		SMEs	
		Number	% of column *N*	Number	% of column *N*	Number	% of column *N*
WTP	Total	260	100.0	347	100.0	309	100.0
	No	49	18.8	17	4.9	19	6.1
	Yes	211	81.2	330	95.1	290	93.9

Tourism Offers and Tourist Taxes

If the notion of positive WTP refers to tourism businesses that claim to be willing to pay at least one of the six proposed tax instruments, there exists a high level of commitment among Andalusian tourist businesses (Table 12.5). This is shown by the high percentage of businesses that are willing to contribute to raising the public revenue required to finance the development of policies related to the competitiveness and sustainability of the destination; more specifically 81.2% of sole traders, 95.1% of micro-enterprises, and 93.9% of SMEs. The chi-squared test with a p-value of 0.000 indicates that WTP differs by segment, being lower among sole traders (81%). The WTP expressed by micro-enterprises and SMEs does not differ significantly.

If WTP is analyzed globally, with respect to the six proposed tax instruments, the degree of acceptance of each of these instruments varies significantly (Table 12.6). The declared WTP was highest for the following tax instruments: tax on activities in protected areas (81.1%), environmental conservation tax in municipalities whose main activity is eco-tourism (72.5%), tax on passenger boats in protected natural areas (71.7%), tax on overnight stays at P2P accommodation (70%), and facility usage charge (61.8%). Only the waste tax payable by tourist businesses had an acceptance of less than 50% (46.8%).

The chi-squared ratio comparison tests with their p-values of 0.000 for the six figures indicate that there are significant differences in WTP depending on the segment. Post-hoc comparisons indicate that:

Table 12.6 Willingness to pay: proposed tax instruments

		Total		Sole traders		Micro-enterprises		SMEs	
		N	%	*N*	%	*N*	%	*N*	%
Tax on overnight stays at P2P accommodation	Total	916	100.0	260	100.0	347	100.0	309	100.0
	No	275	30.0	123	47.3	97	28.0	55	17.8
	Yes	641	70.0	137	52.7	250	72.0	254	82.2
Waste tax payable by tourism businesses	Total	916	100.0	260	100.0	347	100.0	309	100.0
	No	487	53.2	172	66.2	196	56.5	119	38.5
	Yes	429	46.8	88	33.8	151	43.5	190	61.5
Facility usage charge	Total	916	100.0	260	100.0	347	100.0	309	100.0
	No	350	38.2	134	51.5	130	37.5	86	27.8
	Yes	566	61.8	126	48.5	217	62.5	223	72.2
Tax on passenger boats in protected natural areas	Total	916	100.0	260	100.0	347	100.0	309	100.0
	No	259	28.3	102	39.2	85	24.5	72	23.3
	Yes	657	71.7	158	60.8	262	75.5	237	76.7
Tax on activities in protected areas	Total	916	100.0	260	100.0	347	100.0	309	100.0
	No	173	18.9	68	26.2	49	14.1	56	18.1
	Yes	743	81.1	192	73.8	298	85.9	253	81.9
Environmental conservation tax in municipalities whose main activity is eco-tourism	Total	916	100.0	260	100.0	347	100.0	309	100.0
	No	252	27.5	97	37.3	87	25.1	68	22.0
	Yes	664	72.5	163	62.7	260	74.9	241	78.0

- Sole traders have a lower WTP than other segments.
- SMEs have a higher WTP than the other segments for the tax on overnight stays at P2P accommodation, facility usage charges, and waste tax payable by tourist businesses.
- With regard to the environmental conservation tax in municipalities whose main activity is eco-tourism and the tax on passenger boats in protected natural areas, SMEs and micro-enterprises have the same level of WTP.
- The tax on activities in protected areas is the most widely accepted instrument among micro-enterprises.

In terms of the number of tax instruments that Andalusian tourist businesses are willing to pay, this also varies by segment (p-value K-W 0.000). The average number of tax figures a business stated it was willing to pay was correlated to the size of the business. Larger companies claim they would be willing to pay a larger number of tax figures (Table 12.7). More specifically, while sole traders said they were willing to pay an average of 3.32 figures out of the six proposed, micro-enterprises said they were willing to pay 4.14, and SMEs 4.52.

Finally, Andalusian tourism businesses were asked whether they considered tourism a core activity for the economy of the area, whether investments should continue to be made, and whether more and better tourist services should be offered, to increase the number of tourists visiting the area (Table 12.8). In general, there is a high degree of agreement regarding the importance of tourist activity in Andalusia (6.05), the need to continue making investments in the territory, and the idea of offering more and better tourist services (6.29). The p-value of the K-W test for both statements is 0.000, so there are significant differences in opinions

Table 12.7 Number of figures per cluster

		Total	Sole traders	Micro-enterprises	SMEs
Figures	*N* Valid	916	260	347	309
	Mean	4.04	3.32	4.14	4.52
	Standard deviation	1.91	2.08	1.66	1.84

Table 12.8 Importance of tourist activity and the need for new investments

		Total	Sole traders	Micro-enterprises	SMEs
Do you feel that tourism is a core activity for the economy of the area?	*N* Valid	916	260	347	309
	Mean	6.05	5.37	6.32	6.32
	Standard deviation	1.14	1.30	0.92	0.96
Do you think that further investments should be made and that more and better tourist services should be offered to increase the number of tourists visiting the area?	*N* Valid	916	260	347	309
	Mean	6.29	6.16	6.45	6.22
	Standard deviation	0.87	0.72	0.83	0.99

between segments. Sole traders consider tourism to be less crucial to the economy in the area and deem further investments to attract tourism to be less necessary.

Conclusions

Policymakers should consider tourist activity as a domain that needs to achieve, among other objectives, the protection of the natural, cultural, and artistic heritage of a destination. This is particularly the case in places where tourism has severe impacts on the territory, as occasionally happens in well-established and popular destinations, which may lead to both the deterioration of the resources of the destination and a decrease in the quality of the tourist experience. In order to advance the sustainable management of tourism with the goal of achieving economic, social, and environmental objectives, in addition to providing a high-quality offer, it is necessary to understand and anticipate the changes that are taking place in the profile of the tourist.

In this regard, market trends indicate a reduction in the importance of unsustainable mass tourism destinations as the new tourist demonstrates a greater awareness and respect for the environment, while also demanding higher quality services. Moreover, the degradation of the local environment resulting from tourism activities has generated an economic decline in many tourist destinations and given rise to the depletion of non-renewable resources. Tourist destinations need to undertake the renewal and maintenance of attractions, as well as to develop policies that improve the competitiveness of the destination, focusing chiefly on sustainability. In the economic theory of production (Sickles & Zelenyuk, 2019), when production or consumption of goods and services create negative externalities not accounted for in the market price, tax imposition can raise social welfare via more efficient utilization of resources and accounting for the external cost.

Our findings provide evidence of certain tax instruments linked to tourism's potential to generate the necessary revenue to develop policies related to competitiveness and sustainability and to allow for the rejuvenation of Andalusia as a leading global tourist destination. Specifically, nine out of every ten businesses surveyed express their WTP at least one of the six proposed tax instruments. The revenue generated by these taxes would enable the development of policies related to improving competitiveness and sustainability and extend the life of Andalusia as a tourist destination.

These results confirm the findings obtained in previous studies verifying the WTP of tourist businesses to improve the sustainability of a destination (Bernard et al., 2009; Lan et al., 2014; Mäntymaa et al., 2019). In addition, the two hypotheses proposed have been validated. First, one or more segments of the tourism offer have manifested different WTP tourism taxes as part of their commitment to making the tourist destination more sustainable and competitive. Second, there are certain tax instruments that are more widely accepted by Andalusian tourism businesses.

Regarding the first hypothesis, three clusters have been identified. The three variables that were most influential in creating these clusters were business size, activity, and turnover. However, not all the businesses that make up these clusters display the same WTP. More specifically, 81.2% of sole traders, 95.1% of micro-enterprises, and 93.9% of SMEs reported they would be willing to pay at least one of the six instruments proposed. The chi-squared test indicates that WTP differs by segment, being lower among sole traders.

With regard to the second hypothesis, the findings confirm that the degree of acceptance of each of these instruments varies significantly among the different clusters. The most widely accepted instruments among Andalusian tourism businesses are clearly linked to the protection of the natural environment. Specifically, these are a tax on activities in protected areas (81.1%), an environmental conservation tax in municipalities whose main activity is eco-tourism (72.5%), and a tax on passenger boats in protected natural areas (71.7%).

Additionally, there are significant differences in the WTP depending on the segment; sole traders are less willing to pay. As regards the number of tax instruments that Andalusian tourism businesses are willing to pay, this also varies by segment, so that the average number of figures that obtained a positive WTP is correlated to the size of the business.

Regarding management, based on the results obtained, the commitment to sustainability and competitiveness by Andalusian tourism businesses, expressed through a majority WTP, is a good starting point for real progress toward the sustainable management of a previously unsustainable mass tourism destination. Contrary to policymakers who only take into account the modality of an instrument that they consider to be fair for taxes and/or fees, it is important to also consider the opinion of the tourist companies as a reference point. This thinking is in line with Lindsey and Holmes (2010) and Mazzocchi et al. (2019), who assert that in order to achieve greater acceptance in the implementation of a tax instrument, policymakers should assess both the WTP of the agents concerned and their preferred form of tax instrument.

Nonetheless, the results obtained do come with some limitations, including the absence of interviews conducted with tourism businesses in the restaurant trade, since this type of tourism business is not required to register with Andalusia's Tourism Register. The short time frame available in which to conduct the interviews was also a challenge. A survey that spanned an entire calendar year would have further benefited the study. Nonetheless, the findings remain valuable as recommendations for the consideration of tourism-related taxation towards sustainbility.

References

Adedoyin, F. F., Seetaram, N., Disegna, M., & Filis, G. (2021). The effect of tourism taxation on international arrivals to a small tourism-dependent economy. *Journal of Travel Research, 0*(0), 1–19. https://doi.org/10.1177/00472875211053658

Aguiló, E., Alegre, J., & Sard, M. (2005). The persistence of the Sun and sand tourism model. *Tourism Management, 26*(2), 219–231. https://doi.org/10.1016/j.tourman.2003.11.004

Banús, S. C. H. (2021). Specifics of the tax system to support tourism before and during the COVID-19 pandemic: Evidence from South American countries. *SHS Web of Conferences, 92*. EDP Sciences. https://doi.org/10.1051/shsconf/20219201016

Benner, M. (2020). The decline of tourist destinations: An evolutionary perspective on overtourism. *Sustainability, 12*(9), 3653. https://doi.org/10.3390/su12093653

Bernard, F., Groot, R. S., & Campos, J. J. (2009). Valuation of tropical forest services and mechanisms to finance their conservation and sustainable use: A case study of Tapantí National Park, Costa Rica. *Forest Policy and Economics, 11*(3), 174–183. https://doi.org/10.1016/j.forpol.2009.02.005

Birdir, S., Ünal, Ö., Birdir, K., & Williams, A. T. (2013). Willingness to pay as an economic instrument for coastal tourism management: Cases from Mersin, Turkey. *Tourism Management, 36*, 279–283. https://doi.org/10.1016/j.tourman.2012.10.020

Butler, R. (1980). The concept of a tourist area cycle of evolution: Implications for management of resources. *Canadian Geographer, 24*(1), 5–12. https://doi.org/10.1111/j.1541-0064.1980.tb00970.x

Chapman, A., & Light, D. (2016). Exploring the tourist destination as a mosaic: The alternative lifecycles of the seaside amusement arcade sector in Britain. *Tourism Management, 52*(1), 254–263. https://doi.org/10.1016/j.tourman.2015.06.020

Chiu, T., Fang, D., Chen, J., Wang, Y., & Jeris, C. (2001). A robust and scalable clustering algorithm for mixed type attributes in large database environment. *KDD '01: Proceedings of the Seventh ACM SIGKDD International Conference on Knowledge Discovery and Data Mining*, 263–268. https://doi.org/10.1145/502512.502549

Corbisiero, F., & Monaco, S. (2021). Post-pandemic tourism resilience: Changes in Italians' travel behavior and the possible responses of tourist cities. *Worldwide Hospitality and Tourism Themes, 13*(3), 401–417. https://doi.org/10.1108/WHATT-01-2021-0011

Crespi-Vallbona, M., & Mascarilla-Miró, O. (2021). Street art as a sustainable tool in mature tourist destinations: A case study of Barcelona. *International Journal of Cultural Policy, 27*(4), 422–436. https://doi.org/10.1080/10286632.2020.1792890

Drius, M., Bongiorni, L., Depellegrin, D., Menegon, S., Pugnetti, A., & Stifter, S. (2019). Tackling challenges for Mediterranean sustainable coastal tourism: An ecosystem service perspective. *Science of the Total Environment, 652*, 1302–1317. https://doi.org/10.1016/j.scitotenv.2018.10.121

Gomis-López, J. M., & González-Reverté, F. (2020). Smart tourism sustainability narratives in mature beach destinations. Contrasting the collective imaginary with reality. *Sustainability, 12*(12), 5083. https://doi.org/10.3390/su12125083

Gutiérrez, M. J. M., Torelló, J. C. O., & Sebastián, M. S. (2018). Spain as the scenery of mass tourism phenomena–between elite tourism and popular tourism: The image of the country through cinema and photography. In *The Routledge Handbook of Popular Culture and Tourism* (pp. 340–352). Routledge. https://doi.org/10.4324/9781315559018-30

Instituto de Estadística y Cartografía de Andalucía. (2022). *Encuesta de coyuntura turística de Andalucía.* https://www.juntadeandalucia.es/institutodeestadisticaycartografia/badea/operaciones/consulta/anual/20071?CodOper=b3_271&codConsulta=20071

Instituto Nacional de Estadística. (2022a). *Encuesta de gasto turístico.* Retrieved March 12, 2022, from https://www.ine.es/dyngs/INEbase/es/operacion.htm?c=Estadistica_C&cid=1254736177002&menu=ultiDatos&idp=1254735576863

Instituto Nacional de Estadística. (2022b). *Movimientos turísticos en fronteras.* Retrieved March 12, 2022, from https://www.ine.es/jaxiT3/Tabla.htm?t=23988

Lan, N. H. P., Maeda, Y., & Otsuka, K. (2014). Willingness to pay for biodiesel fuel for tourist boats in Halong Bay, Vietnam. Paper presented at the Oceans 2014 MTS/IEEE, Taipei, Taiwan, April 7–10. https://doi.org/10.1109/OCEANS-TAIPEI.2014.6964358

Levin, N., & Zahavi, J. (2001). Predictive modeling using segmentation. *Journal of Interactive Marketing, 15*(2), 2–22. https://doi.org/10.1002/dir.1007

Lindsey, G., & Holmes, A. (2010). Tourist support for marine protection in Nha Trang, Vietnam. *Journal of Environmental Planning and Management, 45*(4), 461–480. https://doi.org/10.1080/0964056022014350З

López-Sánchez, Y., & Pulido-Fernández, J. I. (2014). Factores determinantes de la DAP por un destino más sostenible. El caso de la costa del sol occidental. *Cuadernos de Turismo, 33*, 199–231.

Mak, J. (2006). Taxation of travel and tourism. In L. Dwyer & P. Forsyth (Eds.), *International handbook on the economics of tourism* (pp. 251–265). Edward Elgar Publishing.

Mansourian, S., & Dudley, N. (2008). *Public funds to protected areas.* WWF International. https://doi.org/10.13140/RG.2.1.3869.5286

Mäntymaa, E., Tyrväinen, L., Juutinen, A., & Kurttila, M. (2019). Importance of forest landscape quality for companies operating in nature tourism areas. *Land Use Policy, 107*, 1–12. https://doi.org/10.1016/j.landusepol.2019.104095

Mazzocchi, C., Sali, G., & Ruggeri, G. (2019). Tourists' preferences for alpine pastures maintenance. *Landscape Online, 68*, 1–18. https://doi.org/10.3097/LO.201968

Nelson, K. M., Partelow, S., Stäbler, M., Graci, S., & Fujitani, M. (2021). Tourist willingness to pay for local green hotel certification. *Plos ONE, 16*(2), e0245953. https://doi.org/10.1371/journal.pone.0245953

Pérez, C. (2011). *Técnicas de segmentación. Conceptos, herramientas y aplicaciones.* Gaceta Grupo Editorial.

Pulido, J. I., & Cárdenas, P. J. (2012). La política turística en el ámbito local: la gestión pública de destinos. In J. I. Pulido-Fernández (Ed.), *Política Económica del Turismo* (pp. 213–232). Pirámide.

Sickles, R. C., & Zelenyuk, V. (2019). *Measurement of productivity and efficiency: Theory and practice.* Cambridge University Press. https://doi.org/10.1017/9781139565981

Simeoni, F., Cassia, F., & Ugolini, M. (2019). Understanding the weak signals of demand in a mature tourist destination: The contribution of a sustainable approach. *Journal of Cleaner Production, 219*, 775–785. https://doi.org/10.1016/j.jclepro.2019.02.072

United Nations World Tourism Organization. (2021, January 28). *2020: The worst year in the history of tourism, with a billion fewer international arrivals.* https://webunwto.s3.eu-west-1.amazonaws.com/s3fs-public/2021-01/210128-barometer-es.pdf?afhE7NpuFgX_3avC5b8GTiE2T7Ptcw9J

United Nations World Tourism Organization. (n.d.). *Sustainable Development* [Press release]. https://www.unwto.org/sustainable-development
Wilson, S., Fesenmaier, D., Fesenmaier, J., & Van Es, J. (2001). Factors for success in rural tourism development. *Journal of Travel Research, 40*(2), 132–138. https://doi.org/10.1177/004728750104000203

13

Walking in Bodh Gaya: A Gendered Approach to the Analysis of Sustainable Tourism Policies

Deepika Chandra Verma, Meenakshi Prasad, and Janardan Krishna Yadav

Introduction

UNESCO declared Bodh Gaya, India, a World Heritage Site in 2002 (Sarao, 2020). It is the place where Buddhism originated and it remains a site of spiritual and historical importance to Buddhists and Hindus (Geary et al., 2017). Bodh Gaya is a religious tourism destination and pilgrims come from all parts of the world to witness the places where

We are using the term "gendered" in a limited sense and confining our focus on women.

D. C. Verma (✉)
Indian Institute of Management Calcutta, Kolkata, India
e-mail: deepika11@iimcal.ac.in

M. Prasad
PG Department, Geography, Magadh University, Bodh Gaya, India

J. K. Yadav
Jindal Global University, Sonipat, India
e-mail: jkyadav@jgu.edu.in

T. Walker et al. (eds.), *Sustainable Tourism*,
https://doi.org/10.1007/978-3-031-43528-7_13

Fig. 13.1 Mahabodhi temple (inside view). (Source: Picture taken during our data collection work in Bodh Gaya in 2018)

Buddha's life events unfolded, including the Mahabodhi tree where he achieved Nirvana (see Fig. 13.1). In Bodh Gaya, like many other rustic historic towns, urban development occurs on a piecemeal basis while locals attempt to preserve the city's authenticity (Rodriguez, 2017). Sustainability in tourism activities in such destinations is always a challenge as large infrastructure projects, the broadening of roads, the modernizing of transport, and other civic amenities may take away from the destination's authenticity and make it look like any other urban town. Walking tourism is a popular way to experience a destination as it allows tourists to immerse themselves in local culture and nature as well as interact with locals. Walking tourism can be a sustainable and easily developed option with very little investment required to help achieve a more interactive experience. After all, spiritual walking tourism or pilgrimage has played an important role in laying the foundation for sustainable tourist destinations; the Wakayama pilgrimage trail in Japan, another Buddhist spiritual tourism destination, exemplifies this (Kato & Progano, 2017). Bodh Gaya is perhaps the holiest of places for many Buddhist spiritual travelers, and understanding its walkability is important for academic literature, tourists, and local urban management organizations.

The World Tourism Organization (2019) notes that fostering walking tourism is a great way to boost regional development. It has many benefits, such as it being easy to create, requiring little investment, having significant market potential, creating local economic benefits, and having a high potential for the well-being of the residents. The United Nations World Tourism Organization (UNWTO) suggests some key variables to keep in mind while developing a walking tourism destination, namely (World Tourism Organization, 2019):

- Route characteristics (attractiveness, safety, level of difficulty, access);
- Necessary facilities (signage, toilets, benches, etc.);
- Maintenance system;
- Economic opportunities (accommodation, transport, retail, cultural programs, links to gastronomy and wine tourism, etc.); and,
- Marketing and commercialization.

Existing tourism literature highlights additional factors affecting walkability, such as accessibility, comfort, traffic safety, crime security, and connectivity (Vural Arslan et al., 2018), alongside landscaping, maintained sidewalks, the absence of highspeed and heavy traffic, visual interest, and a sense of place (Lo, 2009).

The walkability of a city is important from the viewpoint of public health as well, as it is related to a lesser dependence on vehicles, which is related to obesity, nutrition, and air quality (Vural Arslan et al., 2018). Pedestrian experiences and expectations are not only important for tourism bodies but also for urban planning in developing countries like India, where cities are expanding and smaller towns are growing into cities.

Walking around a city is perhaps the most natural way to explore it and has not changed much over time. Flâneurs (as poet Charles Baudelaire calls them) are common not just as tourists but also as locals admiring life in the city. Yet, the term remains the "masculine noun" (Wearing & Wearing, 1996). We follow Wearing and Foley (2017), who argue that the literature on the tourist experiences of cities lacks studies from a feminist perspective and that a gendered approach can provide a more realistic and nuanced understanding of the issues and experiences of walking in cities. There is a need to understand the tourist experience and city perspectives from a feminist perspective. We maintain that the tourist-host dichotomy unnecessarily creates a divide in knowledge of city experiences. As Wearing and Foley (2017) note:

> by deconstructing the host-guest dualism we can accommodate the complex web of host-guest relationships that exist in contemporary cities; including, for example, diverse occupational structures where workers are brought in to the city from other areas or are themselves travelers, and the diversity of tourism types where we may have a volunteer tourist who lives in the city but stays elsewhere in that city whilst volunteering.

Studies focusing on tourists' urban walking experiences are rare, but as Farkić et al. (2015) suggest, we too argue that the walking experiences of locals can provide a great deal of insight for city authorities and urban planning agencies, which can benefit locals and tourists alike. We hold that if a local finds a destination hard to navigate on foot, the same will be true for visitors and tourists. Often, urban walking may reveal answers to questions related to the pedestrianization of a destination as well as understanding how locals see and understand their city. This can unravel the "existence of alternatives ways of a re-engagement with city values and history" and how locals "explore the city by themselves and re-trace their proper itineraries" (Kanellopoulou, 2018). Urban walking, as Kanellopoulou (2018) suggests for the historical city of Athens, juggles tourism, social exchange, and acts of citizenship; the same is also true for Bodh Gaya.

A place's social value is the result of the diverse people who engage with it and the resultant social exchanges that occur between those populating the space. In their study of Chinese, Thai, and Indonesian incentive tours to Sydney, Edwards and Foley argue that visitors seek immersive experiences in the city rather than a tour arranged for them to simply use the tourist gaze; they define these as "experiences in which they [tourists] can actively participate, rather than be a spectator" (Edwards & Foley, 2015: 21). Thus, tourism goes beyond city sights and becomes socially embedded. We are cognizant of the larger reach of depicting the "tourist as a chorister"[1] (Wearing & Wearing, 1996) but want to problematize policy moves that do not improve walkability, which can lead to further problems for tourists and host populations rather than enabling inclusive and sustainable tourism.

Ferguson (2011) demonstrates that while tourism development may, in theory, contribute to gender equality and women's empowerment, a substantive reframing of policies is necessary to maximize this potential. In this study, using Nilsson et al.'s (2016) framework, we analyze data related to gendered walking experiences on the acculturation of women walkers who have moved to the walking locale because of an employment opportunity at the local university. Literature on walking and other forms of slow tourism has explored the relationship between tourism and development in policymaking circles (Becken, 2019; Matiku et al., 2020). Sustainable tourism stands for making optimal use of environmental resources that constitute a key element in tourism development, maintaining essential ecological processes, and helping to conserve natural heritage and biodiversity (UNWTO, n.d.). Thus, this kind of tourism implies that the socio-cultural authenticity of host communities (cultural heritage and traditional values) must be preserved, and their built environment must also be conserved. Sustainable tourism also implies, long-term socio-economic benefits to all stakeholders should be fairly distributed to contribute to poverty alleviation and economic growth. Achieving sustainable tourism is a continuous process and requires

[1] In the work of Wearing and Wearing (1996), the tourist destination is represented as "chora," an interactive space. Here, the tourist "becomes a creative, interacting 'choraster' who takes home an experience which impacts on the self in some way." Both the tourist and the host move beyond the dominant hegemonic culture into something akin to another space or what Wearing and Wearing (2006) call "Third Spaces."

constant monitoring on the part of the policymaker while maintaining a high level of tourist satisfaction by providing them with a meaningful experience (UNWTO, n.d.).

Research Context and Methodology

Bodh Gaya and Its Position as a Tourism Destination

Bodh Gaya has long been a site of religious relevance and tourism (Geary, 2008). The Mahabodhi temple, managed by the Bodhgaya Temple Management Act (1949), was recognized as a World Heritage site by UNESCO in 2002 (Sarao, 2020). This recognition has increased the tourist footfall in Bodh Gaya from 6.1 million in 2001 to 35 million in 2019 (Rani, 2020). While trade and commerce have flourished due to this increase in tourism (Agrawal et al., 2010), policy intervention to embellish and remodel many areas in the community (Marketing Division, 2003; PTI, 2021) to improve the experience of tourists has become an issue. These enhancements, in many instances, have jeopardized local ways of life and livelihoods.

Some of these changes include a cap on the number of shops and villages within the 2-km buffer zone proposed in the City Development Plan (Geary, 2008), a restriction on new constructions and the height of buildings within this buffer zone (Department of Tourism, 2002), and special permission for the construction of hotels, resorts, and lodges (authors' own anecdotal findings). The City Development Plan also restricts vehicle entry from Node 1 (a gateway to community resources like restaurants and shopping areas and to the Mahabodhi temple and other assorted temples) except for e-rickshaws to improve walkability during the peak tourist season, from October to March (Qadir, 2018b) (Refer to Fig. 13.2).

The ban on the entry of vehicles was first enforced in February 2018 by the traffic department. Per this plan, the entry of all vehicles was barred from Node 1 onward for environmental and security concerns. Visitors must disembark at Node 1 and either walk the one-kilometer distance

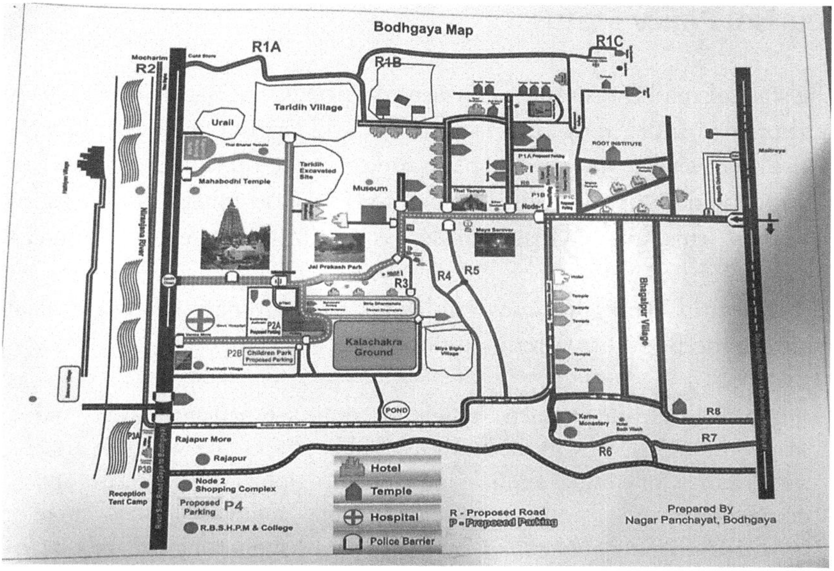

Fig. 13.2 Map of Bodh Gaya. (Source: Nagar Panchayat, Bodhgaya)

between Node 1 and the main shrine or hire an e-rickshaw to cover the distance. For this purpose, the administration allows a limited number of e-rickshaws to ferry passengers. This vehicle ban was severely criticized by locals and was followed by a three-day-long strike (Focus Group Discussion; Qadir, 2018b).

This traffic plan was modified in October 2018 after the Chief Minister's visit for a Rs 145 crore art convention facility project. The relaxation applies to specific vehicles such as tourist buses and private cars to escort passengers up to the Royal Thai monastery (about 350 meters from the main shrine) and was put into effect on November 1, 2018 (Qadir, 2018c). In addition, our focus group discussion reveals that administrative vehicles, vehicles carrying the temple monks and dignitaries, and vehicles escorting tourists staying at hotels located around this area are also allowed to ferry passengers. However, until and unless parked within the hotel/temple premises, these vehicles are expected to return to Node 1 for parking purposes. This modified arrangement has failed to satisfy many locals (Qadir, 2018c).

Tourism Policy Status

The state of road transport has improved since 2003 (National Highways Authority of India, n.d.), but facilities are still lagging (authors' observations; interviews). The current plans are to create the Buddhist Tourism Circuit[2] by developing Bodh Gaya as a year-round tourism destination. However, this puts great stresses on infrastructure development (PTI, 2021).

The World Bank proposes guidelines to rejuvenate the pilgrimage routes in the Buddhist Circuit and develop,

> sustainable tourism products, aiming to connect experiences, assets and spaces through a network of multi-use public spaces and walkable areas with a variety of cultural and religious activities along the routes, sites and destinations and encourage site-sustainable solutions to promote pedestrian-friendly public space environments by supporting safe walking, meditation, gathering and recreation while reducing carbon emissions. (Abakerli et al., n.d.)

The policy intervention seen in Bodh Gaya is more of an attempt to convert shared pedestrian streets to transit pedestrian streets[3] where the conversion improves walkability at a much lower cost than installing any additional expensive infrastructure. Yet, a push for walking does not necessarily result in enhanced walkability; as our findings show, this push may actually be counter-productive, emphasizing one aspect of sustainable forms of movement at the expense of others (See Fig. 13.3).

---

[2] The Buddhist Circuit is a route tracing the footsteps of "Buddha from Lumbini in Nepal where he was born, through Bodh Gaya in Bihar, India where he attained enlightenment, to Sarnath and Kushinagar in Uttar Pradesh, India, where he gave his first sermon and attained Mahaparinirvana. Four additional sites in the region are linked to some of Buddha's most significant life events: Rajgir and Vaishali in Bihar and Sravasti and Sankasia in Uttar Pradesh. These eight sites constitute an 'inner circle' of the Buddhist Circuit, from which Buddhism" spread (Abakerli et al., n.d.).

[3] Pedestrianized streets are categorized into three types—traditional pedestrian streets, shared pedestrian streets, and transit pedestrian streets (Akyuz & Zengin, 2012, as cited in Vural Arslan et al., 2018). Traditional pedestrian streets prohibit all types of vehicles, shared pedestrian streets allow both pedestrians and low-speed vehicle traffic, while transit pedestrian streets only allow public transport in addition to pedestrians.

Fig. 13.3 Walking in Bodh Gaya. (Source: Picture taken during our data collection work in Bodh Gaya in 2018)

Method and Data

The data for this study is sourced from focus group discussions followed by interviews with female faculty members of Magadh University in Bodh Gaya. We conducted 12 interviews with these participants. We used purposive sampling and snowballing and focused on female faculty members who joined Magadh University (which is in the locale) in 2018 and currently live in Bodh Gaya. We chose these women because of their ability to compare their past experiences and any changes in their perception after the policy change. We also purposefully included female faculty members who joined a year later in our sample to understand their experiences and perception of the policy measure, of which they had no prior experience. In addition, we collected data from local newspapers to collate the impact of this policy intervention on other subaltern groups.

Our focus groups encompass female faculty members for two reasons; first, the local context of Bodh Gaya has largely remained patriarchal. As a new generation of women enter the professional field as university professors, mostly hailing from urban spaces, they perceive unique differences between urban landscapes and Bodh Goya and take particular note of the absence of women during evening walks or jaunts. While they may have been used to *flânerie* where they lived before Bodh Gaya, the social context of the examined environment now limits that freedom. While walking is not part of the local culture, the next best option, driving, becomes limited for women during the tourist season in some locations, as there are again limited parking facilities near the area prohibited for vehicles. The second reason for targeting these women is that as newcomers in 2018, their walking experiences would be starkly different from their recent experiences, three years later (at the time of data collection in 2021), by which time they had spent enough time (for acculturation) in Bodh Gaya to be considered locals.

The data analysis was ongoing and iterative, with simultaneous analyses of secondary data (mostly from newspaper articles and published reports by the Tourism Department), interviews, and notes taken during the focus group discussions to support the emergent design, which we regularly compared to see if saturation had been reached or not (Strauss & Corbin, 1990). One co-author was solely responsible for collecting the data, informing the participants about our objective, and obtaining their prior consent for data collection. We anonymized the names of the respondents to maintain privacy (Table 13.1).

Framework

Nilsson et al. (2016) highlight that while the SDGs are interlinked, policymaking and planning occur in silos. Therefore, they propose a framework that provides a systematic rubric to map these interactions easily to help policymakers create "coherent policies and strategies" that go beyond synergies and trade-offs, identifying allies and those who require negotiation. We use this framework which categorizes SDGs interactions into a seven-point scale called "goals scoring" (Nilsson et al., 2016). The aim is

Table 13.1 Key participants

Sl. No.	Codes assigned to participants	Age approx.	Year of joining	Gender	Occupation
1	Anita	36	2019	Female	Assistant Professor, Magadh University
2	Chhaya	35	2018	Female	Assistant Professor, Magadh University
3	Aruna	35	2019	Female	Assistant Professor, Magadh University
4	Malini	44	2018	Female	Assistant Professor, Magadh University
5	Miya	35	2019	Female	Assistant Professor, Magadh University
6	Priya	37	2019	Female	Assistant Professor, Magadh University
7	Sindhu	42	2018	Female	Assistant Professor, Magadh University
8	Sagun	35	2018	Female	Assistant Professor, Magadh University
9	Shivani	35	2018	Female	Assistant Professor, Magadh University
10	Harsha	36	2018	Female	Assistant Professor, Magadh University
11	Smriti	37	2018	Female	Assistant Professor, Magadh University

to aid policymakers and researchers in identifying and testing pathways which minimize negative impacts. The most positive interactions are assigned a score of +3 and the most negative ones, a score of –3. According to Nilsson et al. (2016), these goals can be applied to study the interaction effect of goals, individual policies, or actions.

The positive interactions aid in building synergies, while the three negative interactions are subject to trade-offs. However, in a geographical context, timescale and spatial scale are relevant in categorizing actions/policies/targets/goals, and the framework does not look to generalize. Differences in geography, resources, governance, and technology may lead to different revelations. Bioenergy may counteract food security in heavily populated countries, whereas it increases farmers' and forest owners' resilience in the Nordic region, for instance. Timescale and spatial scale matter as well. Industrial development that acts as an engine of growth may cause pollution and adversely affect the local environment and people's health, while in the long run, this same industrial development can reduce resource sustainability or overuse of commons on one hand while also generating wealth and better health infrastructure on the other.

We use this framework for two reasons: first, because the interactions framework is intuitive, easy to use, and broadly replicable and, second, because it mirrors the fact that SDGs are interlinked and changes in one SDG can lead to changes in another (Nilsson et al., 2016). This, in many instances, implies that we either need to capture synergies or make trade-offs between achieving one SDG over another. Therefore, the framework helps assess and evaluate the governing bodies policies or actions; our chapter applies this framework with respect to governing body actions to encourage sustainable tourism in Bodh Gaya.

Findings and Analysis

Two major themes emerged in our data analysis: (1) the need to improve the ambience for visitors and (2) the presence of increased issues for subaltern participants, for example, local women walkers, the elderly and the ill, cab or taxi drivers, and local business-owners. The themes emerged from our discursive analysis of the data collected from the interviews, focus group discussions, and newspaper articles.

Improving Ambience for Visitors

Reduction of Traffic and Increase in Safety and Experience for Tourists

The new traffic plan that came into effect on February 19, 2018, prohibited all vehicles from entry at Node 1 up to Node 2, covering a distance of about one kilometer to the west and north of the Mahabodhi temple. People could walk or hire e-rickshaws from the two Nodes. The city regulators also controlled the movement of e-rickshaws allowing 150 out of 348 registered e-rickshaws in Bodh Gaya to operate on any given day and allowing the others to function on a rotational basis. The terminal point for e-rickshaws was affixed as the Children's Park (approximately 1100 m from Node 1). Conditional passes were to be allocated during a limited duration to monasteries and hotels that were located in the restricted zone, while alternative routes were planned for locals, residents, and businesses.

The initial reasoning for the prohibition of vehicles from Node 1 was to reduce traffic, ensure the safety of tourists, and alleviate pollution concerns. Regulators, for example, struggled to uphold a "no honking zone" around the Mahabodhi temple, this would be easily resolved with a total vehicle prohibition. A respondent living in Bodh Gaya since her appointment at the university in January 2018 claims that "The prohibition during the peak season definitely increase(s) the convenience, safety [and] walkability for the tourists." According to the interviewees, the traffic ban in the Node 1 to Node 2 area has indeed reduced road traffic, making it much easier for walkers to commute. Most of those who come to the Node 1 stretch for religious reasons or ambience find their experience good. However, this experience also depends upon the season the tourists or visitors visit.

Increased Issues for Women and Other Subaltern Participants

The women who came to Bodh Gaya to work at the university found the transition difficult, especially when these changes adversely affected their day-to-day lives. A respondent, Prof MP, states that:

> for locals it is not convenient as they do not have interest in sight-seeing. I especially find it inconvenient for myself. This prohibition was removed during the COVID times due to a low influx of tourists. I found it more convenient.

Another respondent stated, "This was not a problem before lockdown as there was a lot of crowd [sic], e-rickshaws were easy and cheap to get. Now it is a hassle—don't take that route at all."

Safety: Location and the Time of the Day

Many of our respondents stated that they found it much easier to walk during the daytime than in the evenings. This finding is not new and, in many instances, studies on the topic are disproportionately about daytime walkability (Golan et al., 2019). However, the context of this study reveals that while some respondents have not met any difficulties while walking, others have experienced discomfort or lack of safety while doing so. A lack of safety due to the prohibition of driving is a recurrent theme in many interviews:

> Safety concerns are the primary thing that makes me afraid of walking in Bodh Gaya. People here (male walkers) in general have a problem of ogling that makes me uncomfortable. The stretch of road between Node 1 and 2 is quite isolated. At night you can hardly see any local female out on the road. Though cars have been banned after Node 1 two wheelers were not and rash bikers are another of my safety concerns. After 6 in the evening roads become quite isolated except the area near the Mahabodhi Temple. The parking is also isolated and not well lit during the night. I am a diabetic and often need to use loos. Good and clean loos are not available.

Feeling objectified by male walkers, vulnerable to attacks or mugging instances, and aware of the lack of other female walkers make these women walkers feel uncertain and make them more conscious of their surroundings. During a discussion with one of the respondents, we understood that it was the lack of other women walkers that made them reluctant to walk. A lack of security led some to walk in larger groups or use carpooling services. Reports of muggings and petty theft also made the transition to walking while the ban was in effect more difficult, as these only exacerbated feelings that security was lacking:

> Recently I have heard about a couple of snatching incidences which has once again made me wary of walking. If I, as a person who is native to the state and half local [acculturation], am now afraid then how will I convince other females regarding the safety of walking in Bodh Gaya. Perhaps this is a problem faced by women in most places.

> I find it safer to walk during the day. In the evenings, because of [the lesser] number of walkers and traffic, I find it unsafe to step out. Possibly if we have more police or security posted around this discomfort will get reduced.

Loss of Freedom and Ease

Most of our interviewees felt that shifting to Bodh Gaya had restricted their movement, especially for those women who prefer to drive rather than walk and for those who need to take a car because they have to cross the National Highway to reach the Bodh Gaya Node 1 area. Due to weather conditions, walking is also inconvenient because it is sometimes too hot to walk comfortably. Further, as the major shopping centers are in the banned area for vehicles, shopping is also affected.

> I shifted to Bodh Gaya in January 2018 after joining Magadh University as an Assistant Professor…The place is too hot for the major part of the year which makes walking cumbersome. After taking my car to Bodh Gaya it became my preferred mode of transport, but there is a restriction in Bodh Gaya, i.e., before the lockdown you could not take your car beyond a fixed point referred to as Node 1. However, when I had first come to Bodh Gaya for document verification in December 2017 cars could easily go up till Node 2 which is comparatively closer to the Mahabodhi Temple. The major shopping places were near and beyond Node 2, so if I was able to go up till there by driving then it would have been more convenient for me.

Walkability Questionable for Locals

Aesthetics and the ability to walk with ease are vital to encouraging walking (World Tourism Organization, 2019). However, the presence of individuals seeking money can make walking uncomfortable for some, especially for women walkers. Women are more often targeted for their vulnerability and are also significantly more prone to being a victim of crime, which can also make walking uncomfortable in many circumstances (Dobbs et al., 2009; Vural Arslan et al., 2018).

> There are a lot of [people seeking money on the streets] in Bodh Gaya. This also makes me uncomfortable while walking. [...] As per my observation after the lockdown crime has increased as the police has become laxed in the absence of foreign tourists. But don't the locals and Indian tourists matter?

Reduction in Tourist Traffic and Tourism-Related Revenue

While this change has affected the daily lives of these women, the regulatory intervention also proved detrimental to tourism-related activities. Many tourists rescheduled their trips to other locations because they were having problems with limited access or discomfort due to the restricted traffic. Local businesses and the elderly are increasingly negatively impacted by the restrictions placed on Node 1. In an article published by the *Times of India* in February 2018 (Qadir, 2018a), Bodh Gaya Nagrik Manch president Suresh Singh states that "elderly tourist[s] will be strongly discouraged by the new arrangement. Local customers too would now be forced to walk as only few e-rickshaws have been allowed to ply." Further, local stakeholders express their discomfort with the regulatory intervention:

> Asked about the future plan of action, Dharmendra Yadav, a local stakeholder said a decision on the issue will be taken at the general body meeting of all people affected by the new arrangement. Local business, particularly the hospitality sector, suffered huge losses and a sizeable number of foreign tourists rescheduled their plan making an early departure. According to Arun Kumar Ojha, a prominent tour organizer, domestic tourists have deserted Bodh Gaya. A sizeable number of foreign tourists too have made an exit and gone to places like, Nalanda, Rajgir[,] and Varanasi. It is bound to have a long-term effect on tourist inflow also. (Qadir, 2018b)

Putting these findings into the seven-point goals scoring frame, we find that in some instances, walking and walkability facilitate the possibility of making tourism sustainable (as shown in interactions of levels +3 to +1). In other cases, even with improved walkability and inclusive solutions, various participants/actors in this domain will find it difficult to adjust to the enforced changes (as demonstrated in the interactions of levels −1 to −3).

Discussion and Conclusion

We use the SDG Interactions Framework to understand the contradictory goals of sustainable tourism. Focusing on SDG interlinkages helps us explain this issue on two levels. First, applying the SDG interactions in the research context aids in a systemic framing of policies and highlights the failure of top-down government actions. Second, it helps uncover the contrary dialogues of tourism promotion and sustainability that conflict with each other as well as with the material implications for women walkers that only unfold at the micro-level and are mostly neglected. These findings question the level of inclusivity associated with tourism promotion and sustainable tourism practices on the ground. Our research adds to policy implications for female stakeholders in several important ways. We show that even policies or actions taken by governing bodies at the local level can be as top-down as those made at the central/union level in the policy action to push walking in the area beyond Node 1. Our focus on SDG interactions helped us position sustainable tourism in the policy domain as a space that is much more complex than it appears to be because of possible synergies being canceled out by unexpected trade-offs or repercussions. We argue that working toward sustainable development through some targets on environmental and health-related discourses can lead to issues in the social or gendered contexts. Our findings show that assumptions regarding traffic management and environmental well-being fail to acknowledge the complexities of the behavior of women walkers that are informed by much more than the possibility of crime; our findings show that, for many, their behavior is deeply rooted in social identities and pragmatic action.

Urban planning and architecture literature suggests that walkability is important for the quality of life of the locals, for sustainable and prosperous communities, as well as promoting the leisure and urban tourism markets. Farkić et al. (2015) studied Novi Sad in Serbia and Koper in Slovenia and they suggest that proper walkability is necessary for both tourists and locals though their purposes may differ. If a city is comfortably walkable from a local's point of view, this provides an adequate environment for sustainable walking tourism in the city. Our findings suggest that while it is much easier for able-bodied, male tourists, and male locals to transition from driving or commuting to walking, the experiences of women or (and) disabled and older local walkers may differ. The issues related to enforced walking transcend infrastructure improvements that improve walkability, for example, well-lit, clean, wide pathways and greenery as well as a sense of safety, hospitability as well as connectivity to parks, public amenities, and history and cultural areas of the city (Farkić et al., 2015), especially for women walkers. The women walker often becomes objectified, an observation which is not new but noteworthy for policymakers. The transition to a time when women flâneuse will have the same insights and experiences as those of the male flâneur will require society to change its mindset. Our work therefore shows that administratively enforced changes which shift the way the city is experienced can worsen the experience of subaltern groups. We therefore find that sometimes measures taken to improve sustainable tourism are not inclusive for diverse groups and thus become counter-productive. Our study extends Nilsson et al.'s (2016) conclusion that policymakers "need up-to-date empirical knowledge on how the goals and interventions of one sector affect another positively or negatively" by adding that in some cases, steps or interventions taken to improve some targets within the purview of an SDG can lead to affecting other targets within the purview of the same SDG adversely or facilitate it (as shown in Table 13.2).

Thus, our study suggests that the current policy intervention, which forces patrons to walk in the area beyond Node 1, although fundamentally encouraging more sustainable tourism practices, also has negative implications. Walking boasts many benefits, it presents an opportunity to reduce resource use as well as encourage responsible consumption and positive climate actions. It can also increase the number of patrons who may choose to walk, especially women, which can prove beneficial for gender equality and help women feel safer and more accepted in their environment. However, this policy, which encourages walking, also limits the work of cab-drivers, hurting the interest of local cab-drivers. It also increases the possibility of crimes against women especially during certain times of the day or in areas less frequented by travelers or locals and adversely affects the elderly, the ill, and the disabled from experiencing the city and its sites.

Table 13.2 Using the SDG interactions framework in the context of walking as a policy intervention for sustainable tourism

Interaction	Name	Explanation	In reference to walking and sustainable tourism
(1)	**(2)**	**(3)**	**(4)**
(+)3	Indivisible	It is difficult to separate the achievement of one goal from another	Sustainable tourism by encouraging walking will inevitably reduce resource use leading to responsible consumption and climate action (Goal 12)
(+)2	Reinforcing	Leads to achieving another goal	Sustainable tourism by encouraging walking will eventually lead to an improvement in city/community infrastructure like streets, pavements, facilities, etc. (Goal 8)
(+)1	Enabling	Facilitates another goal	Increased walkers on the road, especially women may encourage opportunities for local women to walk rather than take other fuel-using modes of transportation, enabling gender equality (Goal 5)
0	Consistent	No interactions	Encouraging walking does not significantly affect an increase in income and employment of the local community (Goal 8)
(-)1	Constraining	Limits the achievement of another goal	Increased walking can constrain plying of cabs and other commercial vehicles hurting the locals who see it as an opportunity to earn from local tourism avenues (Goal 8, 11)

(*continued*)

Table 13.2 (continued)

Interaction	Name	Explanation	In reference to walking and sustainable tourism
(-)2	Counteracting	Counters the another goal	A push to walking can lead to an increase in the crime against women especially during certain times during the day or in areas less frequented by travelers or locals (Goal 5)
(-)3	Cancelling	Another goal's achievement is made impossible	Increased push to walking cannot be enforced on the physically disabled, the elderly, and/or infirm travelers or locals (Goal 11)

Source: Adapted from Nilsson et al. (2016)

Notes: Goal 5: Gender Equality, Goal 8: Decent work and economic growth, Goal 11: Sustainable cities and communities, Goal 12: Responsible consumption and production

Our findings show that there is a gender divide that needs to be addressed first to make acceptable recommendations in sustainable tourism, improving subaltern participants' experience. This divide is not only cultural but also, in many senses, universal, where women do not feel safe or feel objectified. These findings show that context would play a very large role in providing acceptable and inclusive solutions for attaining sustainability in tourism. Our work extends Rodriguez's (2017) findings.

Since the Union Ministry of Tourism plans to develop Bodh Gaya as a major Buddhist pilgrimage site, it plans to set up the required infrastructure and facilities to position it as a year-round tourism destination (PTI, 2021). However, as we have seen, small policy interventions can also lead to many complications; policymakers can draft policies by considering SDG interactions to make their efforts inclusive and sustainable.

In summary, SDG interlinkage aids us in offering insights into conflicting discourses and practices toward sustainable tourism. Furthermore, it provides us with a framework to identify the limitations of policy intervention for sustainable tourism. Our study raises a key issue that policymakers need to address while looking for tourism-related solutions toward sustainability: a systematic analysis of policy interventions and their impacts on the key actors. We present a small-scaled case study of policy intervention and change in Bodh Gaya to draw these insights, and in the future, other popular practices for sustainable tourism may also be assessed to understand their implications on subaltern groups. A push toward walking does not necessarily improve walkability, which could have been in theory, a key enabler of sustainable tourism.

References

Abakerli, S., Chamberlain, A., Seth, S., Saxena, S., Khatri, N., & Kuypers, C. (n.d.). The Buddhist Circuit in South Asia. World Bank. https://documents1.worldbank.org/curated/en/387491563440124268/pdf/The-Buddhist-Circuit-A-Program-for-the-Development-of-the-Buddhist-Circuit-in-South-Asia.pdf

Agrawal, M., Choudhary, H., & Tripathi, G. (2010). Enhancing Buddhist tourism in India: An exploratory study. *Worldwide Hospitality and Tourism Themes, 2*(5), 477–493. https://doi.org/10.1108/17554211011090102

Akyuz, M. and Zengin, B. (2012), Tarihi Alanların Yayalaştırılmasının Turist Memnuniyetine Etkileri: Sultanahmet Meydanı Örneği (The effect of the pedestrianization of historic areas on tourist satisfaction: The sample of Sultanahmet Square), *International Journal of Social and Economic Sciences 2*(2), 89–94.

Becken, S. (2019). Decarbonising tourism: Mission impossible? *Tourism Recreation Research, 44*(4), 419–433. https://doi.org/10.1080/02508281.2019.1598042

Department of Tourism. (2002). Information Dossier for nomination of Mahabodhi Temple Complex, Bodhgaya as a World Heritage Site. Ministry of Tourism and & Culture, Government of India. https://whc.unesco.org/uploads/nominations/1056rev.pdf

Dobbs, R. R., Waid, C. A., & Shelley, T. O. C. (2009). Explaining fear of crime as fear of rape among college females: An examination of multiple campuses in the United States. *International Journal of Social Inquiry, 2*(2), 105–122.

Edwards, D., & Foley, C. (2015). Asian incentive events in New South Wales: Expenditure and retail impact. Submitted to Business Events Sydney. https://opus.lib.uts.edu.au/bitstream/10453/36914/1/asian_incentive_events_nsw.pdf

Farkić, J., Perić, D., Lesjak, M., & Petelin, M. (2015). Urban walking: Perspectives of locals and tourists. *Geographica Pannonica, 19*(4), 212–222. https://doi.org/10.5937/geopan1504212f

Ferguson, L. (2011). Promoting gender equality and empowering women? Tourism and the third Millennium Development Goal. *Current Issues in Tourism, 14*(3), 235–249. https://doi.org/10.1080/13683500.2011.555522

Geary, D. (2008). Destination enlightenment: Branding Buddhism and spiritual tourism in Bodhgaya. *Bihar. Anthropology Today, 24*(3), 11–14. https://doi.org/10.1111/j.1467-8322.2008.00584.x

Geary, D., Sivaramakrishnan, K., Kaimal, P., & Yang, A. A. (2017). *The rebirth of Bodh Gaya: Buddhism and the making of a world heritage site (Global South Asia)*. University of Washington Press.

Golan, Y., Wilkinson, N., Henderson, J. M., & Weverka, A. (2019). Gendered walkability: Building a daytime walkability index for women. *Journal of Transport and Land Use, 12*(1). https://doi.org/10.5198/jtlu.2019.1472

Kanellopoulou, D. (2018). Walking the public: Re(visiting) Athens's historical centre. *International Journal of Tourism Cities, 4*(3), 298–315. https://doi.org/10.1108/ijtc-12-2017-0084

Kato, K., & Progano, R. N. (2017). Spiritual (walking) tourism as a foundation for sustainable destination development: Kumano-kodo pilgrimage, Wakayama, Japan. *Tourism Management Perspectives, 24*, 243–251. https://doi.org/10.1016/j.tmp.2017.07.017

Lo, R. H. (2009). Walkability: What is it? *Journal of Urbanism: International Research on Placemaking and Urban Sustainability, 2*(2), 145–166. https://doi.org/10.1080/17549170903092867

Marketing Division. (2003). Final report on 20 years perspective tourism plan for the state of Bihar. (Document: 2002026/UP/FR). Department of Tourism Ministry of Tourism and Culture, Government of India. https://tourism.gov.in/sites/default/files/2020-04/Bihar.pdf

Matiku, S. M., Zuwarimwe, J., & Tshipala, N. (2020). Sustainable tourism planning and management for sustainable livelihoods. *Development Southern Africa, 38*(4), 524–538. https://doi.org/10.1080/0376835x.2020.1801386

National Highways Authority of India. (n.d.). Bihar building new avenues of progress. Ministry of Road Transport & Highways, Government of India. https://morth.nic.in/sites/default/files/PragatiKiNayiGati/bihar/files/assets/basic-html/page-1.html#

Nilsson, M., Griggs, D., & Visbeck, M. (2016). Policy: Map the interactions between Sustainable Development Goals. *Nature, 534*(7607), 320–322. https://doi.org/10.1038/534320a

PTI. (2021, October 7). Tourism Ministry plans to develop Bodh Gaya as round-the-year tourist destination. *The Hindu*. https://www.thehindu.com/

news/national/other-states/tourism-ministry-plans-to-develop-bodh-gaya-as-round-the-year-tourist-destination/article36874344.ece

Qadir, A. (2018a, February 21). New traffic plan for Bodh Gaya to ensure shrine security: SSP. *The Times of India*. https://timesofindia.indiatimes.com/city/gaya/new-traffic-plan-for-bodh-gaya-to-ensure-shrine-security-ssp/articleshow/63003546.cms

Qadir, A. (2018b, February 26). 3-day Bodh Gaya bandh against traffic ban ends. *The Times of India*. https://timesofindia.indiatimes.com/city/patna/3-day-bodh-gaya-bandh-against-traffic-ban-ends/articleshow/63070507.cms

Qadir, A. (2018c, October 29). New Bodh Gaya traffic plan from November 1. *The Times of India*. https://timesofindia.indiatimes.com/city/patna/new-bodh-gaya-traffic-plan-from-november-1/articleshow/66405449.cms

Rani, A. (2020, February 8). 2019 saw highest number of tourists in Bihar. *The Times of India*. https://timesofindia.indiatimes.com/city/patna/2019-saw-highest-number-of-tourists/articleshow/74013593.cms

Rodriguez, J. (2017). Cleaning up Bodhgaya: Conflicts over development and the worlding of Buddhism. *City & Society, 29*(1), 59–81. https://doi.org/10.1111/ciso.12112

Sarao, K. T. S. (2020). The history of Mahabodhi Temple at Bodh Gaya. https://doi.org/10.1007/978-981-15-8067-3

Strauss, A., & Corbin, J. (1990). *Basics of qualitative research: Grounded theory procedures and techniques* (2nd ed.). SAGE Publications, Inc.

UNWTO. (n.d.). Sustainable development. https://www.unwto.org/sustainable-development

Vural Arslan, T., Durak, S., Dizdar Gebesce, F., & Balcik, B. (2018). Assessment of factors influencing walkability in shopping streets of tourism cities: Case of Bursa. *Turkey. International Journal of Tourism Cities, 4*(3), 330–341. https://doi.org/10.1108/ijtc-11-2017-0071

Wearing, B., & Wearing, S. (1996). Refocussing the tourist experience: The flaneur and the choraster. *Leisure studies, 15*(4), 229–243. https://doi.org/10.1080/026143696375530

Wearing, S., & Wearing, M. (2006). "Rereading the Subjugating Tourist" in neoliberalism: Postcolonial otherness and the tourist experience. *Tourism Analysis, 11*(2), 145–162. https://doi.org/10.3727/108354206778001512

Wearing, S. L., & Foley, C. (2017). Understanding the tourist experience of cities. *Annals of Tourism Research, 65*, 97–107. https://doi.org/10.1016/j.annals.2017.05.007

World Tourism Organization. (2019). Walking tourism – Promoting regional development. UNWTO. https://doi.org/10.18111/9789284420346

Index[1]

[1] Note: Page numbers followed by 'n' refer to notes.

T. Walker et al. (eds.), *Sustainable Tourism*,
https://doi.org/10.1007/978-3-031-43528-7

SPRINGER NATURE

GPSR Compliance

The European Union's (EU) General Product Safety Regulation (GPSR) is a set of rules that requires consumer products to be safe and our obligations to ensure this.

If you have any concerns about our products, you can contact us on ProductSafety@springernature.com

In case Publisher is established outside the EU, the EU authorized representative is:

Springer Nature Customer Service Center GmbH
Europaplatz 3
69115 Heidelberg, Germany

The manufacturer's authorised representative in the EU is Springer Nature Customer Service Centre GmbH, Europaplatz 3, 69115 Heidelberg, Germany. If you have any concerns regarding our products, please contact ProductSafety@springernature.com

Printed and bound by CPI Group (UK) Ltd, Croydon, CR0 4YY
01/10/2026
02225903-0010